Ministry of Agriculture, Fisheries and Food

Plant physiological disorders

Reference Book **223**

London. Her Majesty's Stationery Office

ISBN 0 11 242707 3

Acknowledgements

Photographs were provided by East Malling Research Station to illustrate the following apple disorders:

Cox disease
Leaf spot
Crinkle disorder
Fruit cracking
Freezing injury (pear)
Ribbon scald
External CO_2 injury
Alcohol damage (Cox)
Senescent breakdown (Bramley)

The Glasshouse Crops Research Institute to illustrate:

Lettuce-tipburn
Tomato-blossom end rot
Tomato-boxy fruit
Tomato-greenback

The National Vegetable Research Station to illustrate:

Potato — blackening after cooking
Sprouts — internal browning
Cabbage — vein streak
Cauliflower — scorch
Cauliflower — stem rot.

Contents

Introduction

Physiological disorders occur in all crop plants. The frequency and severity of occurrence can be very variable, depending on a range of external factors.

This Reference Book describes plant physiological disorders that have been encountered by ADAS advisers. Some of these disorders occur commonly and cause considerable losses in yield or crop quality on a widespread basis, whilst others occur rarely, but still may cause appreciable losses in individual cases.

The review has been prepared by the ADAS Plant Physiology Committee and is intended to provide a brief guide to the adviser, farmer, grower and the trade of many, often puzzling, disorders. Discussion on plant physiological disorders is often limited due to confusion over which of the many, sometimes superficially similar disorders is being considered. One of the aims of the review is to provide descriptions and illustrations of the disorders so as to aid and stimulate discussion on plant physiological topics amongst farmers, growers and their advisers.

The review outlines the occurrence and importance of each disorder, describes symptoms and where possible indicates the factors causing or influencing the disorder. The review is not intended to provide comprehensive information about any one disorder. Where detailed published information is available a list of references is given for each disorder. However, published information is sometimes sparse or non-existent. In these cases, particular reliance has been placed on ADAS advisory experience and, where appropriate, results of ADAS experiments have been used as background material.

The advice and comment of various workers from the Agricultural Research Council and universities who have contributed to and commented on the publication are gratefully acknowledged.

R W Swain
ADAS,
Bristol

Cereals

Shrivelled grain or poor finishing in cereals (Plate 1.1)

Summary

Low individual grain weight reduces cereal yields. This type of grain has been referred to as crinkled, poorly finished or shrivelled. A number of factors are involved but the most important are moisture stress, trace element deficiency and various diseases.

Symptoms

Shrivelled grain or poor finishing refers to grains which have developed normally up to the milky-ripe stage and subsequently shrink from their original size or fail to fill the endosperm properly so that the pericarp assumes a crinkled appearance.

It has proved difficult to devise simple tests for the diagnosis of samples apart from visual assessment. The standard grain sieving procedure does not separate shrivelled from normal grain of the same size. Determination of 1000 grain weight does not differentiate between small and plump grain and shrivelled grain of the same weight.

Occurrence

The problem of poor finishing in cereals has been recorded throughout the United Kingdom but no assessment of the area involved has been made. It varies considerably from year to year. Winter wheat appears to be the most susceptible and in bad years up to 30 per cent of crops may be affected. Grain shrivelling reduces both yield and quality and occurs in both low and high yielding fields.

Causes

Shrivelled grain results from the failure of metabolites to reach the grain or the loss of carbohydrates in the grain by excess respiration, the former is the most important. Failure of sufficient metabolites to reach the grain may be caused by inadequate synthesis of substrate in the leaves, peduncle and glumes or by failure to translocate the substrate to the developing grain. Shading experiments by Nosberger and Thorne (1965) showed that restricting the production of metabolites in the vegetative parts of the plant resulted in grain shrivelling. Other factors such as drought, disease and trace element deficiency can also limit metabolite production.

Translocation experiments using radioactive isotopes have shown that there is little movement of metabolites from the grain to the rest of the plant. Water is also lost from the grain as they shrivel and during normal ripening the dry matter of the grain increases due to an increase in carbohydrates and loss of water. Under conditions of moisture stress the supply of metabolites will be reduced as moisture is lost from the grain. Maximum dry matter content of the grain is reached at around 30 per cent moisture content. Observations have show that there is a loss of about 5 per cent of the dry matter during normal ripening due to respiration but that these respiration losses will be higher when the upper leaves and glumes are invaded by parasitic and saprophytic fungi.

Water stress

The proportion of grain shrivelling as a result of water stress may be inversely related to rainfall in June and July especially on soils where there has been a restriction to root development. Skinner (1969) showed that the stage of growth at which water stress occurs is important in determining whether grain shrivelling takes place. Stress at the milk ripe stage was the most important factor in wheat and barley. Moisture stress pre-anthesis increased the proportion of blind spikelets at the base and tip of the ear in wheat and barley. In extreme drought years such as 1976 there is evidence of movement of reserves from the stem to the grain 'sink' (Plant Breeding Institute, Annual Report 1977) but in more normal years the products of pre-anthesis photosynthesis make little contribution to grain filling.

Trace element deficiencies

Copper deficiency has been clearly defined as a major factor in shrivelled grain in the UK (Pizer *et al* 1966, Caldwell 1971, Davies *et al* 1971) especially on soils with a high organic matter content, high pH and high moisture content. Wheat and barley are the most sensitive to copper deficiency, whilst oats are less sensitive. Manganese deficiency occurs on similar soils but is less directly associated with grain shrivelling.

Disease factors

Diseases which damage the area of green tissue late in the growing season will affect grain quality. Shrivelled grain in barley is often associated with late attacks of foliar diseases such as brown rust (*Puccinia hordei*), leaf blotch (*Rhyncosporium secalis*) and mildew (*Erysiphe graminis*). In wheat shrivelled grain is often associated with glume blotch (*Septoria nodorum*), leaf spot (*Septoria tritici*), yellow rust (*Puccinia striiformis*) or mildew. Root and stem base diseases can also seriously affect grain filling and their effects are accentuated under poor soil conditions and drought. Diseased plants have higher rates of respiration.

Other factors influencing grain filling

Cereal crops receiving high levels of nitrogen have large surface areas for transpiration and removal of soil moisture which may be increased by as much as 30 per cent compared with crops receiving lower levels of nitrogen (Rothamsted Experimental Station Reports 1976–77). Such crops are more vulnerable to moisture stress during grain filling. Similarly, high levels of nitrogen increase the proportion of the respiring non-productive parts of the plants, mutual shading of the leaves is also increased so that a higher proportion of the leaf area is below the compensation point. Generally additional nitrogen reduces grain size due partly to a stimulation of ear density and grains per ear. Weather conditions account for most of the variation between years in terms of grain size in wheat and barley.

High temperatures during grain filling increase losses from photorespiration and dark respiration, and under most conditions high temperatures are associated with moisture stress.

Poor grain finishing in wheat is common in most years on organic soils and is attributed to loose soil, poor root development, high soil nitrogen levels or trace element deficiency.

References

Caldwell, T H (1971) Copper deficiency in peats and sands in East Anglia. London, MAFF. Technical Bulletin 21, pp. 73–87.

Davies, D B, Hooper, L J, Charlesworth, R R, Little, R C, Evans, C and Wilkinson, B (1971) Copper disorders in cereals grown on chalk soils in south eastern and southern England. London, MAFF. Technical Bulletin 21, pp. 88–118.

Nosberger, J and Thorne, G N (1965). *Annals of Bot.* **29**, pp. 635–44.

Pizer, N H, Caldwell, T H, Burgess, G R and Jones, J L O (1966) Investigations into copper deficiency in crops in East Anglia. *J. agric Sci. Camb.* **46**, pp. 303–14.

Skinner, R J (1969) Water stress in barley and wheat. *NAAS East Midland Reg. Rept. of Field Expts.* pp. 66–8.

Swain, R W and Melville, S C (1973) Shrivelled grain or poor finishing of cereals. *ADAS Quart. Rev.* **11**, pp. 118–27.

Seedling abnormalities (Plate 1.2)

Summary

Seed with impaired embryos produce seedlings of low vigour and plant loss may be high under adverse seedbed conditions and deep drilling.

Symptoms

Partial germination, failure to emerge and slow emergence of seedlings making them more susceptible to damage from soil pests. Split coleoptiles which are unlikely to produce viable plants, and distorted and coiled first leaves.

Causes

There are many causes of seedling abnormalities including bad harvesting conditions, too rapid drying of wet grain and the use of excessive drying temperatures. Damp grain stored in large heaps leads to suffocation of the embryo. Split grains and those badly chipped by the combine allow entry of fungi and mites resulting in low germination and vigour. Herbicide residues in the soil such as propyzamide and trifluralin can cause distinctive seedling abnormalities.

Experimental

Some seed firms are undertaking vigour testing of seed lots and the test is useful in detecting those with adequate germination but low vigour. ADAS experience with comparisons of high and low vigour seed has shown little advantage under good seedbed conditions.

Herbicide effects (Plate 1.3)

Summary

Growth regulator herbicides can cause ear distortion and loss of yield when applied outside the correct crop growth stages.

Symptoms

In wheat-opposite and supernumerary spikelets, rat tailing and blind spikelets. In barley-whorled and extended rachis, tubular flag leaf trapping the emerging ear, kinked stem and blind spikelets.

Causes

Growth regulator herbicides based on benzoic and phenoxyacetic acid derivatives when applied before GS30 cause ear distortion symptoms, loss in yield is generally rare but the rachis may become brittle and allow shedding in barley. Herbicide application after 'jointing' usually results in severe loss in yield due to a reduction in the number of fertile florets but generally no reduction in grain size.

Frost damage (Plates 1.4 and 1.5)

Symptoms of frost damage can take a number of forms. Direct freezing may be severe enough to kill plants particularly winter oats. The 'Rugby stocking' effect is often seen as yellow and purple markings on emerging coleoptiles. Low temperatures can also affect the develoment of ear primordia in spring barley causing blind spikelets.

Winter barley infected with mildew appears to be more susceptible to winter kill than healthy plants and autumn fungicide applications are sometimes beneficial. Manganese and potassium deficient plants are more susceptible to frost injury.

Frost heave of the soil can break off emerging coleoptiles and produce symptoms similar to pest damage. Barley and oats seem to be more susceptible than wheat to roots breaking at the crown and whole plants being lifted from the soil.

Flag leaf necrosis/Chlorosis in wheat

This condition, sometimes called physiological stress, is common in most years soon after ear emergence, it may be confused with the small speckles present in the cultivars Cappelle, Mega and Maris Huntsman which seem to be genetic. The larger lesions cannot be explained by mineral deficiency or disease.

They are also associated with crops under stress that have received high rates of nitrogen, fungicide and growth regulator. The condition requires further investigation.

Oats — blasting

Summary

Blasting of oats results from adverse environmental conditions between the time of spikelet initiation and panicle emergence. The failure of grain formation in oats can also be caused by pest and disease attacks.

Symptoms

In the crop there are a proportion of spikelets which appear white and little or no grain is formed.

Occurrence

This disorder is very common and a few blasted panicles are present in most crops. It is worse in some years than in others but there does not appear to be any simple correlation with weather conditions. Yield reductions of 10–20 per cent are possible in severe cases.

Causes

Frit fly attack is one cause, but blasting also occurs in the absence of the pest. In the absence of frit fly blasting appears to be due to environmental conditions during the period from the initiation of the spikelets until just before the panicle emerges. A reduced water supply, frost, physical damage and defoliation all increase the incidence. Blasting is more common in late sown crops, and late tillers are apt to show more. Barley Yellow Dwarf Virus is a major cause of blasting.

References

Empson, D W (1958) 'Blast' produced by the artificial defoliation of oats. *Plant Path.* **7**, pp. 85–7.
Sheals, J G (1950) Observations on blindness in oats. *Ann. app. Biol.* **37**, pp. 397–406.

Wheat — loose ear (Plate 1.6)

Summary

Loose ear of wheat is a rare non-parasitic condition which is unlikely to be of much economic importance. The exact cause is uncertain but a number of cases have followed grazing of wheat by stray cattle or sheep.

Symptoms

Bleached white ears, conspicuous amongst green ears, occur in single or in small groups of plants. One or more tillers are affected per plant. The bleached area includes the ear and stem immediately below, whilst the sheath, flag leaf and rest of the plant remain green. The bleached area can be pulled out easily from the sheath and is fractured a few centimetres or so from the topmost joint. There is also a groove extending lengthwise from the base of the ear to point of severance. Following severance, axillary shoots can develop within the sheath from the top node.

Occurrence

This disorder has been reported occasionally but less so in recent years. The number of ears affected in a crop is usually small and doubtlessly yield compensations take place on unaffected tillers and adjacent unaffected plants. Loss of yield in individual cases is small and nationally insignificant.

Causes

In some instances farmers have been unaware that cattle have strayed into the crop though there is evidence of this from hoof marks and droppings. Similar symptoms have been found in wheat demonstration plots caused by frequent passage of demonstrators through the crop.

Reference

Taylor, R E and Schofield, E B (1956) An observation on loose ear of wheat. *Plant Path.* **5**, p. 94.

Grass

Grass — sod pulling (Plate 2.1)

Summary

The pulling-out of sods from grass swards can be a severe problem on some farms. Several causal factors can be associated with the problem and frequently it is difficult to identify a single main factor.

Symptoms

Poor anchorage of grass roots or shallow-rooting of grass permitting sods to be lifted out by grazing animals. These sods can be as large as 30 cm across and 5 to 7 cm deep.

Occurrence

Sod-pulling is of isolated occurrence throughout Britain on a range of soil types. It is often linked with aerial tillering but should be considered as a distinct disorder. Surveys of farms in the West Midlands and Cumbria showed that the occurrence of both aerial tillering and sod-pulling was widespread but that the majority of cases were not serious.

Causes

This problem has been associated with high nitrogen application, low phosphate levels down the soil profile, low pH, poor drainage and soil compaction. Sod-pulling can be aggravated by the resistance of the grazing animal to completely defoliate herbage which has become too mature. Close examination of the botanical composition of a pulled sod sometimes shows that it is composed of inferior grasses which have established in bare patches caused by dung pats.

Investigational work

Work at the Grassland Research Institute, Hurley, using a continuous flow solution has shown that large nitrogen dressings can increase the nitrate nitrogen in the soil to a level which affects grass roots. Thus at 2000 mg/1 nitrate-nitrogen in the soil roots were severely stunted with reduced tensile strength and the ratio of root weight to shoot weight was decreased. Shallow subsoiling of permanent grass at about 22 cm has been shown to be an effective remedial treatment for a problem of severe pulling-out of grass where there was soil compaction at 8 cm. However, shallow sub-soiling should only be undertaken with a machine which does not damage the sward surface. Drastic loosening of the surface soil may, in some situations, aggravate the position by encouraging poor surface rooting particularly if there is adequate moisture available.

Claims that sod-pulling is associated with low soil potassium are unsubstantiated by scientific evidence and it has been observed at all potassium levels on a range of soil types.

Perennial ryegrass — aerial tillering

Summary

Poorly utilised ryegrass in a highly fertile situation can be stimulated into aerial tillering possibly by the establishment of ideal microclimatic conditions in the sward. Improved management of stock, including their nutrient requirements, or programmed topping of the grass will prevent the problem.

Symptoms

Plants develop a stoloniferous habit of growth above ground level.

Occurrence

Aerial tillering is found in areas of poor grass utilisation associated with low stocking rates, and in old dung pats. Its frequency of occurrence is increasing in intensively used grassland and is widespread in swards grown under high fertility conditions but with lax grazing. Suitable microclimatic conditions within the sward encourage aerial tillering. Slight poaching can influence the occurrence.

The problem can largely be overcome by proper management of the grazing requirements of the animals, particularly close grazing in spring.

Reference

Simmons, R G, Davies, A and Troughton, A J (1974) The effect of cutting heights and mulching on aerial tillering in two contrasting genotypes of perennial ryegrass. *J. agric. Sci. Camb.* **83**, pp. 267–73.

Root crops

Carrot — cavity spot

Summary

Cavity spot is widespread in the United Kingdom and causes considerable losses especially in late lifted maincrop carrots. There are large variations in severity between different locations and different years. There is considerable controversy over the cause of cavity spot. Calcium supply or more recently bacterial pathogens have been suggested but there is no conclusive evidence.

Symptoms

Cavity spot was first described and named in America by Guba *et al* in 1961. The first signs which can be seen on fresh roots that have been carefully washed clean of soil, are sunken elliptical depressions up to 5 mm wide on the side of the roots. The skin is intact, the depression resulting from a collapse of the underlying tissue. There is no marked change of colour at the lesion in the early stages. The lesions enlarge as the roots mature and the skin ruptures, often leaving a ragged edge at the border of the cavity. Fungi and bacteria present in the soil may then colonise the exposed root tissue, increasing still further the area and depth of the lesion. However, cavity enlargement by secondary pathogens does not always occur. Suberised lesions are common. Lesions may extend halfway round the root. As rotting organisms may already be present, affected roots can rot rapidly when they are stored after lifting. Small cork-lined lesions without secondary infection have been observed in mature crops and attributed to 'arrested' cavity spot, although confusion with early carrot root fly damage is possible. It is also possible that cavity spot produces only small initial lesions, and all the subsequent enlargement is due to secondary pathogens.

Anatomical observations at the Royal Botanic Gardens, Kew have indicated the following sequence of events in cavity formation:

(1) The cavity orginates in parenchymatous cells situated several cell layers below the root surface; the cytoplasm of some of these cells becomes coarsely granular.

(2) Tannin-like pigments are deposited in the affected cells.

(3) The nuclei degenerate.

(4) Some meristematic activity may be induced in cells surrounding the cavity.

(5) The cells first affected break down to form a cavity.

(6) Cells on the outer side of the cavity, and those adjacent on other sides, begin to break down.

It has been suggested that lactifers may be involved in cavity initiation. However, the carrot contains such a large numbr of lactifers that cavities invariably cover one or more branches of the system, and it has not been possible to establish that the rupture of lactifers initiates cavity formation. (For information on the anatomy of carrot roots see Esau, 1940).

The growth stage at which cavity spot initiates is unknown but cavities have been detected at the pencil-thick stage in some crops.

Occurrence

Although cavity spot occurs widely in the carrot growing areas of the United Kingdom, there are large variations in its severity between sites and between seasons. Advisory experience has shown that cavity spot is generally increased by wet summers and irrigation.

These factors are strongly associated with the occurrence of cavity spot:

Late lifting
There is gradual increase in the severity of cavity spot in the late autumn and winter.

Early sowing
The incidence and severity of cavity spot increases as the crop matures and early sown crops are likely to be harvested at a more advanced state of maturity than late sown crops. This is because early sowing increases the length of the growing season.

Spacing
Closer spacing increased the incidence.

Variety
Large differences in susceptibility between cultivars have been reported. Amsterdam Forcing types, which are lifted early, are seldom affected. Chantenay types are very susceptible with Autumn King types relatively resistant. However, there is no direct evidence from variety trials on affected sites that Autumn King varieties are more resistant than the Chantenay types. It is true that fewer advisory problems occur with Autumn King but these varieties are generally grown at wide spacings for the ware market and are less likely to be affected.

Carrots are grown mainly on peaty soils in the Fens and Lancashire or loamy sands in Norfolk and the East Riding. Cavity spot occurs on both but is generally more severe on peaty soils. Fields producing severely affected crops in one year have produced normal crops in the following year. Most attemps to produce the disorder experimentally, including use of soil from affected fields, have been unsuccessful.

Causes

Calcium, other nutrients, soil conditions, pathogens, pests and cultural factors have been suggested as being implicated but much of the experimental work has been inconclusive. Some of the results in outline are as follows:

Calcium and other nutrients
Calcium deficiency, either actual or induced by excessive potassium levels has been claimed to cause cavity spot (Maynard *et al* 1961, 1963; Duncan and Parsons 1966). Advisory experience and experimental work in the UK has not confirmed this despite studies involving nutrient solutions, foliar sprays of calcium and translocation studies with Ca-4S.

More recent experimental work has given conflicting results on the involvement of calcium. Perry and Harrison (1979 — I) found that neither calcium nor potassium concentrations in the soil nor their ratios were related to cavity spot in surveys of crops in East Scotland. No consistent response to high concentrations of calcium or potassium were obtained in field and pot experiments.

DeKoch, Hale and Inkson (1980) showed that cavity spot incidence was negatively correlated with nitrate in the carrot tissue and suggested that waterlogging results in the loss of soil nitrate by denitrifaction and leaching, forcing the plants to make use of ammonium nitrogen which antagonises calcium uptake. Scaife, Burton and Turner (1980) correlated cavity spot incidence positively with soil ammonium levels and negatively with pH. The calcium concentration in carrots was related negatively to soil ammonium but not to cavity spot incidence.

Soil conditions

Perry and Harrison (1979 — II) of the Scottish Horticultural Research Institute have found cavity spot most commonly in fields with compacted sandy soils and poor drainage but in years of high incidence it also occurs on loamy soils. Cavity spot was induced in field experiments by irrigating compacted carrot beds to excess in July and August, and in pots sealed with wax and stood in water for five days.

In contrast to Scotland cavity spot is found in East Anglia in excessively drained loamy sands without detectable compaction. Soils from Yorks and Notts which have produced cavity spot are similar in structure to those which are associated with the disorder in Scotland. Perry and Harrison (1979 — I) found that fields with a high incidence of cavity spot had a high soil bulk density and, although the converse was not always true, there was a positive correlation between cavity spot and bulk density.

Pathogens

Perry and Harrison (1977 and 1979 — II) claim that an obligatory anaerobic, pectolytic, spore forming bacterium belonging to the genus Clostridium is pathogenic to carrot roots growing in pots and exposed to temporary anaerobic conditions. The resulting lesions are similar to those of cavity spot found in the field providing the roots are kept dry for at least three weeks after inoculation. However, a period of at least three days anoxia on the root surface at 20 °C must be demonstrated before the bacteria can be confirmed as the cause of the symptoms in the field.

From the 1977 crop in Eastern Region aerobic pectolytic bacteria were isolated from cavities in 44 or 45 specimens.

From the 1978 crop when cavities were examined at an even earlier stage of development such bacteria were recovered from 20 of 27 specimens. Over the two years the commonest organisms were Pseudomonas (86 per cent) usually belonging to Groups IVb (60 per cent), IVa (14 per cent) and less frequently Group II, or other aerobic pectolytic bacteria including Flavobacterium and Erwinia. In both years the identical cavities had been examined for pectolytic anaerobic Clostridium. This organism was recovered from the 1977 crop from only four cavities which on the other hand also contained the far commoner Pseudomonas or Erwinia already mentioned and not at all from the 1978 material.

Earlier Perry and Harrison (1977 and 1979 — II) had implicated pectolytic Clostridium based on the inducement of symptoms by experimental treatments in pots.

The observations in Scotland and ADAS in Eastern England can be reconciled on the basis that Clostridium was a later invader of a lesion already occupied by Pseudomonas or other aerobic bacteria and could become the dominant organism should later anaerobic/anoxic conditions persist from excessive wetness and the activity of aerobic bacteria, which is the most rapid method of depleting dissolved oxygen. Similar Pseudomonas bacteria have been isolated in Eastern Region from lesions on the roots of parsnip and horseradish.

Clostridium is a spore-bearing organism commonly occurring in soil. It is generally not considered to be a pathogen although it can become active in stored crops e.g. potatoes.

Spacing and irrigation

Cavity spot is most prevalent in carrots grown at high population densities which have a high oxygen demand per unit volume of soil, and lesions are initiated in the summer when temperatures favour a high soil respiration rate. This will favour the development of anaerobic conditions especially if soil conditions such as impeded drainage, high bulk density, a soil cap or 'slumped' soil reduce the oxygen diffusion rate at the root surface to below the total biological demand and thus render the microsite anaerobic during periods of high rainfall or irrigation (see section on pathogens claiming a pectolytic obligatory anaerobic Clostridium causes cavity spot). Irrigation has given conflicting results, although it often increases incidence it sometimes gives a decrease,.

Uronic acids

Differences in uronic acid levels have been small, being slightly higher in afflicted tissue.

References

De Koch, P C, Hall, A and Inkson, R H S (1980) *J. Sci. Food Agric.* **31**, p. 839.

Duncan, A A and Parsons, J (1966) Cavity spot of carrots and parsnips investigate. *Oregon Veg. Digest,* **15**, pp. 3–4.

Esau, K (1940) Developmental anatomy of the flesh storage organ of *Daucus carota. Hilgardia,* **13,** pp. 175–206.

Grogan, R G, Zink, F W and Kimble, K A (1961) Pathological anatomy of carrot root scab and some factors affecting its incidence and severity. *Hilgardia,* **31**, pp. 53–68.

Guba, E E, Young, R E and Ui, T (1961) Cavity spot disease of carrots and parsnip roots. *Plant Disease Reporter,* **45**, pp. 102–5.

Maynard, D N, Gersten, B, Vlach, E F and Vernell, H F (1961) The effects of nutrient concentration and calcium levels on the occurrence of carrot cavity spot. *Amer. Soc. Hort. Sci. Proc.* **78**, pp. 339–42.

Maynard, D N and Gentile, A C (1963) The distribution of calcium in cells of the roots of carrot (*Daucus carota* L.) *Physiologia Plantarum,* **16**, pp. 40–3.

Perry, D A (1967) Carrot root disorders. *Agriculture,* **74**, pp. 222–5.

Perry, D A and Harrison, J G (1977) Pectolytic anaerobic bacteria cause symptoms of cavity spot in carrots. *Nature,* **269**, p.509.

Perry, D A and Harrison, J G (1979) Cavity spot of carrots I. Symtomatology and calcium involvement. *Ann. appl. Biol.* **93**, pp. 101–8.

Perry, D A and Harrison, J G (1979) Cavity spot of carrots II. The effect of soil conditions and the role of pectolytic anaerobic bacteria. *Ann. appl. Biol.* **93**, pp. 109–15.

Scaife, M A, Burton, A K and Turner, M K (1980) Cavity spot of carrots – an association with soil ammonium. *Communications in Soil Science and Plant Analysis,* **11**, (6), pp. 621–8.

Carrot — five o'clock shadow

Summary

This disorder, which has an unusal but apt name, was first noted in 1958 and has caused considerable losses in England and Scotland. Five o'clock shadow appears as small, light-to-dark-brown spots scattered on the side of the roots usually only after heat peeling. It is a condition induced by localised boron deficiency in the outer cortex of the root.

Symptoms

The skin of the fresh root is sometimes roughened and difficult to remove in the peeling process but no other symptoms are visible until the roots have been heated in the peeling process. Then they appear as small, light-to-dark-brown spots scattered on the side of the roots. If the roots have been steam peeled, the spots are raised pegs of hardened tissue. After lye and carborundum peeling, the spots are still visible although the root is smooth. The discoloration extends radially towards the core of the root and cannot always be removed by a second run through the peeling machinery. The spots have been seen occasionally on large, old roots or on those that have been stored, but the discoloration always becomes darker after peeling.

Occurrence

This disorder occurs spasmodically and is most severe on sandy soils in dry summers and in early lifted crops. It does not occur on peat soils.

Causes

Shadow is caused by localised boron deficiency and occurs in the absence of any leaf symptoms. In affected roots, the boron concentration is low (less than 20 mg/kg) in the outer 2 mm skin but normal (25–30 mg/kg) in the rest of the root. In unaffected roots there is no radial concentration gradient of boron. Patches of rough grey skin visible after washing but before peeling have sometimes been associated with low boron but often appear in roots with adequate boron and can be mistaken for shadow.

Reference

Perry, D A (1967) Carrot root disorders. *Agriculture*, **74**, pp. 222–5.

Carrot — clayburn

Summary

Clayburn appears as an irregular blackened area on the side of the root after heat peeling.

Symptoms

The only symptoms seen on the fresh root are slightly discoloured scruffy areas, often with shallow, longitudinal cracks, or faintly darkened areas. An intense black to blue-black discoloration develops in the secondary phloem following steam or lye peeling. The extent of the discoloration varies but may extend halfway round the root.

Occurrence

Clayburn occurs infrequently on heavy land in wet seasons.

Causes

Not known.

Reference

Perry, D A (1979) Carrot root disorders. *Agriculture,* **74**, pp. 222–5.

Potato — coiled sprout and thick sprout (Plate 3.2)

Summary

Coiled sprout is common in early potato crops. Seed stored in warm conditions with long sprouts and planted into cold soil is the most usual combination of causal factors. Unless severe, yield is only slightly affected.

Symptoms

This disorder occurs principally in early cultivars although in recent years it has been reported to be quite common in maincrops. Instead of emerging normally the sprout forms coils below the surface of the soil. Often the sprouts are thickened and this may occur without coiling. Light brown superficial lesions, usually on the inside of the coil, have been associated with the disorder. The effect is to delay emergence of the sprout. It can be regarded as a less severe form of 'little potato'. Both disorders often occur in the same crop.

Occurrence

Coiled sprout was first investigated in eastern England in 1958 and has been noted in most early potato growing areas since. It is fairly common in early potato crops, particularly where the seed is sprouted. There is some cultivar difference in susceptibility and is more common in vigorous sprouting varieties. Home Guard, Arran Pilot, Arran Comet and Ulster Premier are fairly susceptible but Craigs Royal is less affected. A survey in south west Cornwall in 1968–69 showed 80 per cent of early potato crops affected but only 16 per cent were severely affected (20 per cent plants with coiled sprouts). The disorder is more common in late seasons. The mild form 'thick sprout' is very common and can be found in most crops of first early potatoes.

Causes

The disorder is considered to be due to planting tubers of advanced physiological age into cold soil. The longer or more etiolated sprouts, the earlier the planting and the colder the soil the more severe the disorder. The temperature at which the seed potatoes are stored is the most important factor, high temperatures producing long sprouts. Storage conditions may affect the normal geotropic growth after planting.

Seed taken from early planted crops is more liable to coiling. *Verticillium nubilum* has been shown to cause coiling but in these cases there is no fasciation and coiling caused by *V. nubilum* is probably of little importance. Soil compaction may aggravate the disorder. Green sturdy sprouts with well developed leaf intials are less susceptible.

Importance

Moderate levels (20 per cent plants affected) of the disorder have little effect on earliness or yield. At Rosewarne EHS levels of 30 per cent of coiled sprout have not seriously affected early yield but in severe cases coiling appreciably delays early lifting. The incidence of the disorder can be reduced by lowering the temperature of seed stores with ventilation at night, or by delaying planting. Shallow planting may help to prevent coiling by encouraging earlier emergence. A moderate level of coiled sprout seems likely in all early crops, particularly as the factors that reduce coiling also reduce the early yields of potatoes.

References

Ali, M A, Lennard, J H and Boyd, A E W (1970) Potato coiled sprout in relation to storage treatment and to infection by *Verticillium nubilum*. *Ann. appl. Biol.* **66**, pp. 407–15.

Catchpole, A H and Hellman, J R (1975) Studies of the coiled sprout disorder. *Potato Res.* **18**, pp. 282–9.

Cox, A E (1970) Coiled sprout survey in south west Cornwall, England, in 1968 and 1969. *Potato Res.* **13**, pp. 332–41.

Houghton, B H (1973) Early potatoes studies on the effects of seed storage temperatures. *Expl. Hort.* **25**, pp. 97–101.

Lapwood, D H, Hide, G A and Hurst, J M (1967) An effect of plant soil compaction on the incidence of potato coiled sprout. *Plant Path.* **16**, pp. 61-3.

Moorby, J and McGee, S (1966) Coiled sprout in the potato, the effect of various storage and planting conditions. *Ann. app. Biol.* **58**, pp. 159–70.

Toosey, R D (1965) The influence of sprout development at planting on subsequent growth and yield. In: J D Ivins and F L Milthorpe (Eds) The growth of the potato. London, Butterworth pp. 79–95.

Potato — little potato (Plate 3.3)

Summary

Little potato is moderately common in the earliest crops of potatoes. It occurs when mature seed, that is seed stored in warm conditions, is planted into cold soil. It is associated with coiled sprout disorder and both can occur in the same crop.

Symptoms

The disorder is characterised by the premature formation of small tubers. There is little or no top growth.

Occurrence

Little potato was well known in the Penzance area of Cornwall as early as the beginning of the 19th century. The disorder can be viewed as a more serious form of coiled sprout. It occurs in the same situations but less frequently. On average about one in seven fields are affected and probably less than 5 per cent of the plants in these fields. Cultivars differ in their susceptibility; Arran Pilot and Comet and Ulster Premier are particularly prone.

Pentland Dell is unusual being a maincrop cultivar susceptible to little potato. Late seasons, i.e. those with low spring temperatures, increase the incidence of the disorder, particularly if the winter has been mild giving warm storage conditions. The condition is most likely to occur if the planted tubers were stored in warm conditions (above 9°C) in the dark and were de-sprouted. Little potato and coiled sprout often occur together, but there are cultivar differences in the relative proportions.

Causes

The disorder is due to the planting of physiological mature seed into soil below 8°C. In a warm soil unsprouted seed will grow normally as differentiation of eyes to form stolons does not take place until after planting. Storing at lower temperatures or delaying planting until the soil warms up will help to reduce the incidence. Controlled sprouting under lights helps to reduce little potato. At the same time this may also reduce the early yield so in practice there is always some risk of little potato occurring in first early crops, paticularly in years with low soil temperatures in the spring. Little potato may occur following chilling of the tubers. The fact that little potato is associated with the use of physiologically old seed is well recognised in France where in some districts the planting of certain varieties intended for basic seed production is permitted only after a fixed date in order to prevent that seed getting too old before it is planted the next year. Furthermore certain varieties are down-graded if they are not cool-stored during the winter. Premature tuber formation can occur when the fungus *Rhizoctonia solani* attacks sprouts after planting again more common in cool dry conditions. The lesions on the sprouts and the little potatoes forming on large stolons are symptomatic.

Potato — internal rust spot (IRS) (Plate 3.4)

Summary

This disorder is characterised by rusty spots or blotches found on the outer surface of the tuber. It is suggested that internal rust spot may be produced under conditions of localised calcium deficiency.

Symptoms

The symptoms are seen at normal lifting time. They do not appear to increase in storage or to cause tuber decay. Rusty brown spots or blotches varying in number, size and shape are distributed irregularly over the cut surface of the tuber. It is distinguished from the other types of rusty markings by the shape and position of the marks. In spraing the marks are in narrowish curved streaks and in net necrosis there are spots or fine streaks arranged on both sides of the vascular ring. Potato mop top virus may also produced spraing type symptoms in some cultivars e.g. Arran Pilot but may often be distinguished by one or more raised concentric rings at the tuber surface.

Occurrence

The disorder has been known for many years and is found in many parts of the world. It is sometimes confused with the virus induced conditions spraing, net necrosis and potato mop top and in some cultivars potato acuba. The last named virus is now rare in Britain. It can also be confused with symptoms of chilling.

IRS is only found occasionally. Some cultivars are more susceptible than others. In the South West, Ulster Dale and Arran Consul are the most commonly affected. Maris Peer has also been reported as affected on a number of occasions. It occurs on both light and heavy soils.

Causes

In South Africa a similar disorder, internal brown fleck, was attributed to phosphorous deficiency associated with high acidity (Whitehead *et al* 1953) but Combrink and Hammes (1972) found that phosphorus had no effect on its occurrence, the incidence of which increased with decreasing levels of calcium. Other work (Van Schreven, 1935; Collier and Huntingdon, 1977) has also suggested that internal rust spot (IRS) may be produced under conditions of calcium deficiency.

Collier *et al* (1978) have subsequently shown that tubers of Maris Piper on plants supplied with 3 mM calcium chloride solution developed symptoms of IRS while those plants supplied with 9 mM did not show symptoms and had a substantially greater concentration of calcium in the tuber dry matter. It is suggested that IRS is associated with a low concentration of calcium in the affected tubers.

Further work (Collier *et al* 1979) has shown that in all 10 varieties used, as calcium supplied to the plants increased, the percentage of tubers affected by IRS decreased. The mean incidence of IRS in individual varieties differed considerably but was unrelated to the mean calcium concentration in their tubers. In increasing order of IRS incidence the varieties were as follows: Stormont Enterprise (16.7 per cent), Pentland Dell, Record, Desiree, King Edward, Maris Piper, Pentland Squire, Pentland Crown, Pentland Ivory and Majestic (50.3 per cent).

References

Collier, G F and Huntingdon, Valerie C (1977) Symptoms of calcium deficiency in the potato. Wellesbourne Report of the National Vegetable Research Station for 1976, p. 45.

Collier, G F, Wurr, D C E and Huntingdon, Valerie C (1978) The effect of calcium nutrition on the incidence of internal rust spot in the potato. *J. Agric. Sci. Camb.* **91**, pp. 241–3.

Collier, G F, Huntingdon Valerie C and Wurr, D C E (1979) Internal rust spot in potatoes. Wellesbourne Report of the National Vegetable Research Station for 1977 p. 39.

Combrink, N J J and Hammes, P S (1972) Die invoeld van kalsium, fosfaat en boor op die voorkoms van inwendige brunivelek by aartappels. *Agroplantae,* **4**, pp. 81–6.

Schreven, D A, van (1935) Physiologische proeven met de aardappelplant. *Landbouwkundig tijdschrift,* **47**, pp. 706–26.

Whitehead, T, McIntosh, T P and Findlay, W M (1953) *The Potato in Health and Disease,* pp. 421–6. Edinburgh: Oliver and Boyd.

Potatoes — blackening after cooking

Summary

Blackening is associated with a high ratio in the potato of chlorogenic to citric acid caused by climatic conditions, fertiliser usage or soil moisture conditions. Cultivars differ in susceptibility.

Symptoms

After cooking, the cut surface of the tuber shows a grey discoloration. The colour is more pronounced at the heel end of the tuber and when the potatoes stand after cooking. For this reason it is sometimes called 'stem-end blackening'.

Occurrence

This disorder is related to weather conditions and in some years is fairly common. Cool wet growing seasons are the worst blackening years, and so the disorder is more prevalent in the western districts.

Causes

The blackening of the tuber is caused by a pigment formed from iron and chlorogenic acid during cooking. The degree of blackening depends on the chlorogenic acid content of the tuber and particularly on the chlorogenic acid/citric acid ratio. Factors which affect this ratio are:

Cultivar

Some cultivars e.g. Majestic have a naturally high content of chlorogenic acid. Others e.g. King Edward, Ulster Supreme have a low content and rarely blacken. Pentland Crown, Dell, Hawk, Ivory and Croft are susceptible.

Growing conditions

Cool wet seasons give higher chlorogenic acid contents.

Fertilisers

A low uptake of potassium due to low soil or inadequate application is correlated with low levels of citric acid and more blackening. A high rate of application of ammonium or urea compounds decreases citric acid levels. Nitrate tends to increase the chlorogenic acid content so nitrogenous fertilisers tend to increase blackening. Using potassium sulphate instead of potassium chloride may slightly increase the citric acid content and so be beneficial.

Soil conditions

The disorder is more common on wetter or more moisture retentive soils. In areas where blackening is a problem the recommendation is to grow less susceptible cultivars with the correct level of potassium fertiliser.

References

Burton, W C (1966) The Potato. Wageningen, Veenman, H and Zonen. pp. 193–5.

Evans, J L (1973) A review of NIAB studies on after-cooking blackening in potatoes. *J. Nat. Inst. Agric. Bot.* **13**, pp. 9–20.

Smith, O L (1958) Potato Quality X. Post-harvest treatment to prevent after cooking darkening. *Amer. Potato J.* **35**, p. 373.

Swann, T *et al* (1963) Biochemical aspects of the quality of potatoes. *In:* J D Ivins and F L Milthorpe (Eds.) The growth of the potato. London, Butterworth.

Potatoes — internal bruising (Plate 3.6)

Summary

Blackspot on potatoes is frequently associated with internal bruising of tubers. Their susceptibility is increased largely by high dry matter and low potassium content.

Symptoms

Enzymic browning occurs in the cortex of the tuber following bruising. The zone of the cell damage is termed internal bruising. The condition known as internal rust spot is distinct and maybe of viral origin. (see page 26).

Occurrence

This disorder occurs frequently in susceptible cultivars, such as Record, in which the incidence can be as great as 20 per cent of the crop. It is of particular importance where the crop is used for crisping.

Causes

The extent of internal bruising in a batch of tubers will depend on the susceptibility of the tubers and the mechanical forces applied (see Terrington EHF Annual Reviews for 1971, 1972 and 1976 on the effect of harvester adjustment on internal bruising). Tuber susceptibility has been shown to be dependent on several characteristics including temperature, cell turgidity, dry matter and potassium content (Jacob 1959, Burton 1966). Both increasing dry matter content and decreasing tuber potassium content increase internal bruising susceptibility. Vertregt (1968) showed that internal bruising incidence was low when the tuber potassium was greater than 25 g/kg dry matter. Ophuis *et al* (1968) showed an increasing susceptibility as tuber potassium level decreased; they also showed a linear relationship between increase in dry matter and decrease in internal bruising incidence.

Work by ADAS East Midland Region in 1973 on commercially grown crops showed that of the three cultivars, Record, Pentland Crown and Pentland Ivory, Record was most prone to internal bruising in laboratory bruising tests. Work at Terrington EHF investigating varietal differences in susceptibility to internal bruising between 1972 and 1974 showed susceptibility to decrease in the following order, Pentland Ivory, Majestic, Maris Piper, Pentland Crown, Desiree, King Edward (Terrington EHF Annual Review for 1976 pp. 16–17).

In the East Midlands ADAS work on internal bruising was positively correlated with dry matter content and negatively correlated with tuber potassium content. There was no direct relationship in Record between tuber potassium and soil potassium but there was with Pentland Crown and Ivory. At Terrington EHF three out of four trials between 1972 and 1975 showed that high rates of potassium fertiliser reduce the incidence of internal bruising (see Terrington EHF Annual Review for 1971, 1972 and 1976). Similar effects were observed at the Norfolk Agricultural Station in 1973 and 1974.

References

Burton, W G (1966) The Potato. Wageningen, Veenman, H and Zonen

Jacob, W C (1959) Studies on Internal Blackspot in potatoes. *Mem. Cornell Agric. Exp. Stn.* p. 368.

Ophuis, B G, Hesen, J C and Krorsbergen, E (1958) The influence of temperature during handling on the occurrence of blue discolorations inside potato tuber. *European Potato Journal*, **1**, pp. 48–65.

Terrington EHF Annual Reviews for 1971, 1972, 1973 and 1976.

Vertregt, M (1968) Relations between black spot and composition of the potato tuber. *European Potato Journal*, **11**, pp. 34–44.

Potato — blackheart (Plate 3.7)

Symptoms

The interior cells of the tuber are killed and turn black. This may be followed by rotting. It should be distinguished from 'bruising' where only part of the tuber is affected and from 'blackening' after cooking.

Occurrence

This disorder is uncommon and is only found in exceptional circumstances.

Causes

The prime cause is asphyxiation. This may occur in two situations:
- In tubers in waterlogged soil when some or all of the tubers may be killed.
- In tubers that have been stored in too high a temperature so producing excessive respiration. This can follow when soil cores are formed within the clamp during off-loading into store, when wet tubers are off-loaded into store or when tuber breakdown follows disease. Damage occurs at temperatures at or above 40°C.

Reference

Cox, A E (1967) The Potato. W H and L Collingridge p. 108.

Potatoes — second growth (gemmation, chain tuberisation, hollow heart, cracking, jelly-end rot). (Plate 3.8)

Summary

Second growth is promoted by irregular growth of the potato and possibly high temperatures. There are several forms of this disorder.

Symptoms

Gemmation   is the production of knob-like growths from the eyes, and is the most common form of second growth.

Prolongation of the Rose End   In this form there is a marked waist between the two parts of the tuber.

Chain Tuberisation   Two or three tubers form, separated by short lengths of stolon.

Hollow Heart   A cavity is produced in the centre of the tuber.

Cracking   Deep cracks form in the tuber.

Jelly-End Rot   Starchy material is removed from the heel end of a tuber to build new tissue at the rose end; the heel first becomes translucent or glassy and may afterwards rot in the ground or in storage.

Another form of the disorder is where a tuber breaks dormancy and produces a stolon but no secondary tuber.

Occurrence

Second growth is dependent on growing conditions but in adverse seasons most crops can be affected and as many as 50 per cent of the tubers in the worst affected crops.

The type and extent of second-growth symptoms differ according to cultivar. Majestic commonly shows large cracks and swollen protrusions from the eyes. Golden Wonder is often affected by second growth at the rose end of the tuber. Chain tuberisation is commoner in the round tubered cultivars, Record and Epicure for example. Hollow heart is found in large tubers, such as produced by Arran Banner and Pentland Dell.

Apart from the disadvantages of poor or irregular shape, second growth also affects quality. The transfer of starch from the first formed tuber to the second growth area reduces the dry matter and flavour of the original tubers. Glassy tubers will remain hard after normal cooking. Second growth is less of a problem with modern cultivars.

Irrigation in drought periods reduces the incidence of second growth.

Causes

A period of favourable growing conditions following a check to growth such as rain following a drought can cause symptoms. Recently the suggestion has been made that higher temperatures are the main cause.

References

Burton, W C (1966) The Potato. Wageningen, Veenman H and Zonen, p. 184.
Lyst, C (1960) Second growth phenomena. *European Potato Journal*, **3**, pp. 307–24.

Sugar Beet — bolting (or running to seed)

Summary

Sugar beet is a biennial plant grown as an annual crop and a bolter is therefore any plant which completes or attempts to complete its full growth cycle in a single season. The occurrence of bolting in sugar beet is mainly associated with environmental factors such as cold conditions during the stage when seed is produced in the mother crop, and early sowing of the following crop particularly in conjunction with a cold post-seedling stage.

Symptoms

A bolter is characterised by the production of a seedhead in the year of sowing. It is easily observed in a crop because of its upright habit of growth due to the development of the floral axis in contrast to the flatter appearance of the rosette of leaves of the normal crop. A bolter can be difficult to distinguish in the crop from a bolting weed beet.

A weed beet is a plant produced from any seed present in the soil before the crop was sown and these are a major problem in some beet fields.

Also a further source of weed beet can be sugar beet seed contaminated with annual forms of wild beet. This has occurred occasionally in the past, but sugar beet seed producers now carry out stringent tests to ensure that seed is free from contamination. Besides reducing the yield, bolters may interfere with machine harvesting. Shed seed increases problems of weed beet as seed can survive for 10 years. Field contamination with weed beet is now a serious problem.

Occurrence

Bolting varies greatly from season to season and is usually worse in early sown crops. Cultivated sugar beet is a biennial normally requiring a vernalisation period or cold treatment during the winter to initiate reproductive growth in the following year. A high frequency of bolting can occur in the year of sowing under certain conditions. The tendency towards earlier sowing in combination with the climatic trend towards colder conditions in late spring has been associated with the recent increase in bolting in monogerm cultivars.

Longden *et al* (1975) have proposed three controlling factors:

Environmental effects on the mother seed crop.

Direct environmental factors.

Genetic interaction and environment.

Cold conditions whilst the seed is on the straw can produce seed more susceptible to bolting when sown in the following year. A cold period of 6 to 8 weeks duration at 5 to 10°C will cause vernalisation in the post-seedling phase if the seedlings are in the 4 true leaf stage and if devernalisation is not brought about by warm conditions.

References

Longden, C, Scott, R K and Tyldesley, J (1975) Bolting of sugar beet grown in England. *Outlook on Agriculture*, **8**, (4), pp. 188–93.

Arnold, M H (1979) Weed beet — whose problem: the farmers or the seed producers? *British Sugar Beet Review*, **47**, 1, pp. 5–7.

CEREALS

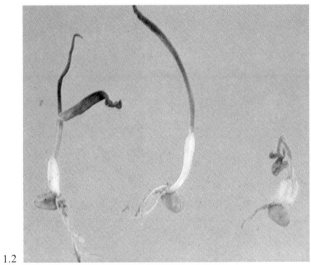

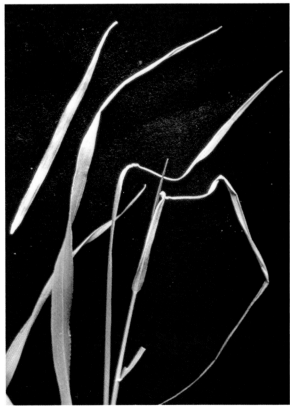

1.1 Shrivelled grain
1.2 Seedling abnormality (propzyamide damage)
1.3 Herbicide effects
1.4 Frost damage (barley)

1.5 Frost damage (barley)
1.6 Loose ear (wheat)

2.1 Sod pulling

ROOT CROPS

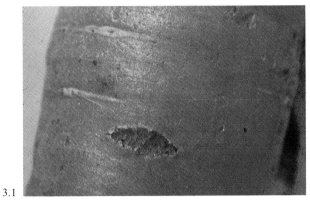

3.1

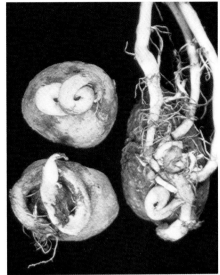

3.2

3.3

CARROT
3.1 Cavity spot
POTATO
3.2 Coiled sprout
3.3 Little potato
3.4 Internal rust spot

3.4

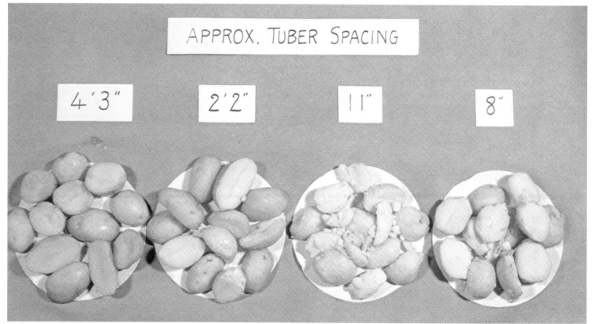

APPROX. TUBER SPACING

4'3"   2'2"   11"   8"

3.5

3.6

POTATO
3.5 Blackening after cooking
3.6 Internal bruising
3.7 Black-heart

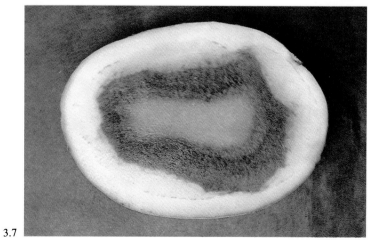

3.7

ROOT CROPS

3.8

POTATO
3.8 Second growth
SUGAR BEET
3.9 Strangles
SWEDE
3.10 Many-necking

3.9

3.10

BRASSICA CROPS

BRUSSELS SPROUTS
4.1 Large necrotic leaf spot (Blackspot)
CABBAGE
4.2 Pepper spot

4.1

4.2

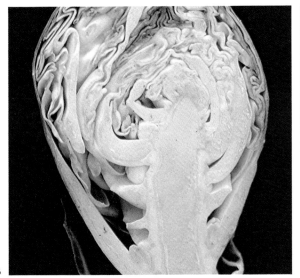

4.3

4.4

CABBAGE
4.3 Internal browning
4.4 Vein streak
4.5 Internal tipburn

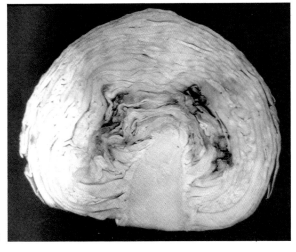

4.5

4.6

CAULIFLOWER
4.6 Scorch
4.7 Winter stem rot (hollow stem)

ROOT CROPS AND BRASSICAS
5.1 Strangles

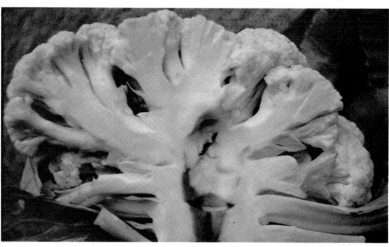

4.7

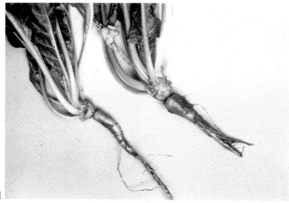

5.1

FRUIT CROPS

6.1

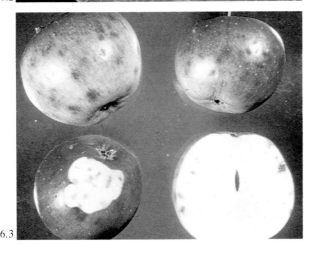

6.2

APPLE
6.1 Cox disease
6.2 Leafspot and leaf drop
6.3 Bitter pit

6.3

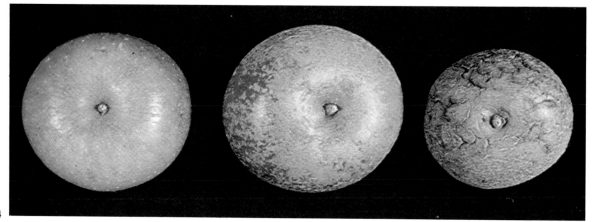

6.4

6.5

6.6

6.7

APPLE

6.4 Cracking and russetting

6.5 Crinkle

6.6 Watercore in Bramley's seedling

6.7 Carbon dioxide injury

6.8

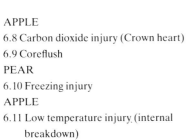

6.9

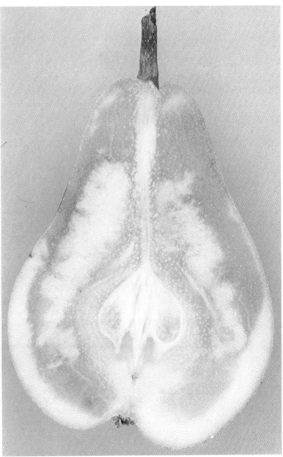

6.10

APPLE
6.8 Carbon dioxide injury (Crown heart)
6.9 Coreflush
PEAR
6.10 Freezing injury
APPLE
6.11 Low temperature injury (internal
    breakdown)

6.11

6.12

6.1

6.14

APPLE
6.12 Low temperature injury (ribbon scald)
6.13 Oxygen deficiency injury
6.14 Senescent breakdown
6.15 Superficial scald

6.15

SALAD CROPS

LETTUCE
7.1 Tipburn
7.2 Tipburn (Glassiness)
TOMATO
7.3 Blossom end rot (BER)

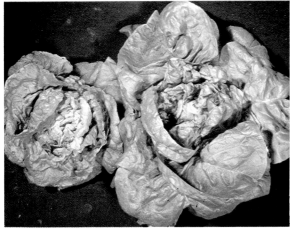

7.1

7.2

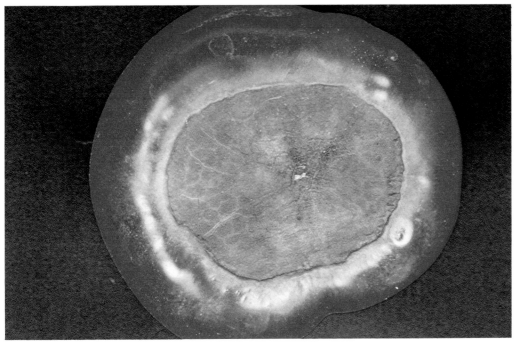

7.3

7.4

7..

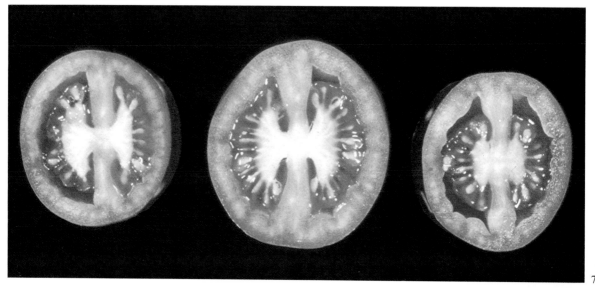

7.6

TOMATO
7.4 Blotchy ripening
7.5 Bronzing
7.6 Boxy fruit
7.7 Greenback

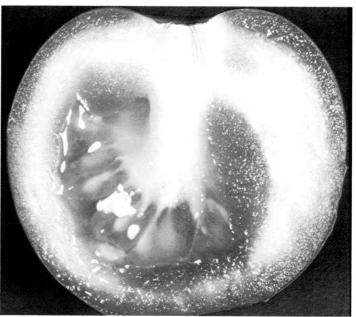

7.7

FLOWER CROPS

ROSE
8.1 Proliferation after budding
NARCISSUS
8.2 Hot water treatment damage

8.1

8.2

TULIP
8.3 Topple
8.4 Bud necrosis
8.5 Blindness

8.3

8.4

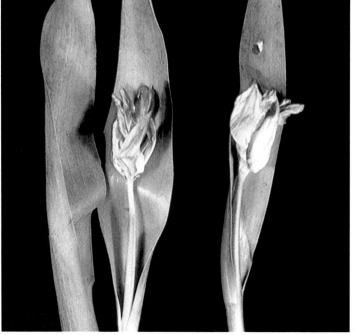

8.5

Sugar Beet — strangles (see page 45) (Plate 3.9)

Swede — cracking

Summary

Irregular growth due to fluctuating moisture or other climatic factors can cause cracking of the root and stem tissue.

Symptoms

Cracking of the root, sometimes deeply, into the flesh. The cracking is often in the shoulder of the root.

Occurrence

The disorder is only found occasionally, but in some cases as much as 5 per cent of the crop may be affected.

Causes

The main cause is a period of good growing conditions following a period of restricted growth. Commonly this is due to damp weather following a spell of dry weather. (Similar effects are found with carrots.)

Swede — many-necking (Plate 3.10)

Summary

This disorder is often associated with pest attack though there are some instances where pests have not been found.

Symptoms

Instead of producing one stem, the main growing point dies and many stems form. The disorder is usually first noticed in July or August, although it may appear earlier in early sown crops in Northern England and Scotland.

Occurrence

Occurrence of the disorder is uncommon and varies from year to year. When it appears, usually no more than 1 or 2 per cent of plants in a crop are affected.

Causes

Many-necking is a reaction to damage to the growing point. Swede midge is one cause but usually only if growing conditions are unfavourable. Cabbage root fly maggots are another cause. If the plants are not actively growing, for example during a period of drought, the maggots can tunnel down from the petiole to kill the growing point. Many-necking due to pest attack can usually be distinguished by the scarred tissue. There have been a few cases of many-necking in the absence of any symptoms of pest damage.

Brassica crops

Brussels sprout — internal browning (Plate 4.3)

Summary

Internal browning is caused by localised calcium deficiency. Some reduction in susceptibility can be achieved by cultural methods, but serious losses still occur.

Symptoms

Internal browning originates as marginal necrosis in the youngest leaves of the sprout. As the leaves expand, the browning appears as a concentric ring of brown necrotic tissue with clearly defined margins. The vascular tissue is usually not affected.

Occurrence

Internal browning is quite common in commercial crops of sprouts, with marked variation in severity between seasons. Since there are no external symptoms, affected sprouts cannot be graded out and quite low levels of incidence can cause the rejection of the whole crop. The incidence seems to be related to sprout size and density and is much higher in large low density sprouts (Faulkner, 1961; Faulkner and Johnson 1963; and Nieuwhof 1971). There are indications that susceptibility to internal browning is an inherited characteristic (Priestley and North 1962). Cultural factors which advisory experience suggest affect susceptibility include temperature fluctuations (Maurer 1964), water stress, root restrictions due to soil compaction or poor drainage, excess nitrogen application (Greenwood and Cleaver 1971), and overmaturity (Carter 1962).

Causes

The calcium concentrations of sprouts is commonly 0.1–0.2 per cent and thus much lower than in the leaves. Internal browning has been induced in sand culture by low calcium levels (Millikan and Hanger, 1966; Millikan et al 1970 and 1971, and Maynard and Baker, 1972). The calcium content in the sprouts decreased from the wrapper leaves to the meristematic region. It is well established that calcium is transported exclusively in the xylem, so storage organs and meristematic tissues have a very restricted supply of calcium. Localised calcium deficiency occurs because of the very low mobility of calcium within the plant, and the competition between leaves and other organs for water and calcium. The supply of calcium to the root is not limiting, since calcium is the dominant cation in temperate soils, and the total calcium content of the plant is greatly in excess of that needed for metabolism, since much of the calcium is immobilised by organic anions.

Plant analysis for total calcium is not always a reliable guide to calcium status. Haworth and Cleaver (1966) found no relationship between sprout calcium and internal browning. In a survey of commercial crops in Lincolnshire in 1973/4 affected crops all had calcium levels below 0.15 per cent. Other experiments have shown the effects of various factors on the incidence of internal browning but it is likely that these act through their influence on the calcium supply of the sprout. In recent experiments in Eire, removing the basal 10–20 leaves when the basal buttons were just visible gave a very large reduction in the incidence of internal browning. De-leafing reduced the proportion of larger sprouts, and the yield of freezing buttons was reduced by about 11 per cent in five out of seven experiments (Anon 1974).

Control

The risk of internal browning can be minimised, but not eliminated, in the following ways.

Irrigate to avoid checks due to water stress, or grow on soils of high available water capacity.

Avoid root restriction due to soil compaction or waterlogging.

Avoid using more nitrogen than needed for maximum yield.

Harvest before the sprouts become overmature.

Try to use cultivars of low susceptibility.

References

Anon, (1974) Horticulture. *An Foras Talentais*, pp. 23–5.

Carter, A R (1962) The stopping of Brussels sprout plants. *Exp. Hort.* **7**, pp. 7–15.

Faulkner, G J (1961) Internal Browning of Brussels sprouts. Wellesbourne, Report of the National Vegetable Research Station for 1960. pp. 15–16.

Faulkner, G J and Johnson, A (1963) Internal browning of Brussels sprouts. *Euphytica*, **12**, pp. 299–310.

Greenwood, D J and Cleaver, T G (1971) Nitrogen fertiliser requirements of Brussels sprouts. *Agric. London*, **78**, pp. 32–5.

Haworth, F and Cleaver, T J (1966) The mineral composition of Brussels sprouts. *J. Sci. Fd. Agric.* **17**, pp. 304–8.

Maurer, A R (1964) A physiological breakdown in Brussels sprouts. *Can. Plant Dis. Survey*, **44**, pp. 265–6.

Maynard, D N and Barber, A V (1972) Internal browning of Brussels sprouts: A calcium deficiency disorder. *J. Amer. Soc. Hort. Sci.* **97**, pp. 789–92.

Millikan, C R and Hanger, B C (1966) Calcium nutrition in relation to the occurrence of internal browning in Brussels sprouts. *Aust. J. Agr. Res*, **17**, pp. 863–74.

Millikan, C R, Bjarason, E N and Hanger, B C (1970) Browning of Brussels sprouts. *J. Agric. Vict.* **68**, p. 312.

Millikan, C R, Bjarason, E N and Hanger, B C (1971) Effect of calcium sprays on the severity of internal browning in Brussels sprouts. *Aust. J. Exp. Agr. Animal Husb.* **11**, pp. 123–8.

Nieuwhof, M (1971) Factors affecting the incidence of internal deviations in Brussels sprouts. *Euphytica*, **20**, pp. 527–35.

Polegaev, V I (1971) Potassium and spotted necrosis damage in cabbage. *Zasc. Rast.* **16**(3), p. 50.

Priestley, W G and North, C (1962) Inheritance of susceptibility to internal browning of Brussels sprouts. *Nature,* **193**, p. 801.

Cabbage — pepper spot (black speck) (Plate 4.2)

Summary

Pepper spot (black speck or spotted necrosis) of cabbage can cause considerable losses in stored winter white cabbage. There are large variations in severity of the disorder between cultivars and sites and there is some evidence that temperature fluctuations, high nitrogen nutrition and the atmospheric conditions of storage can influence the disorder.

Symptoms

Very small superficial black spots less than 1 mm in diameter appear distributed randomly over the leaf surfaces. The spots typically, but not invariably, appear first on the outer leaves of the heads and progress inwards during storage, symptoms are rarely seen in growing crops. In each spot necrosis starts in the stomatal guard cells and spreads to a few surrounding epidermal cells (Geeson and Robinson 1975; Cox 1977).

Similar symptoms have occasionally been seen in related crops, including other types of cabbage and cauliflower.

Occurrence

Symptoms of necrosis on stored white cabbage, apparently identical to those of 'pepper spot' have been described previously as 'grey speck' in the Netherlands (Betzeman and Commandeur 1968, and Nieuwhof *et al* 1974), 'Black Speck' in Florida (Strandberg *et al* 1969), 'Blattpunktnekrosen' in Germany (Theiler 1973) and as 'spotted necrosis' in the USSR (Polegaev 1974). Similar symptoms have also been recorded in other brassicas, e.g. 'blackspot' in spring greens (Cook 1976).

The incidence and severity of pepper spot varies considerably between sites and from season to season, but there is insufficient evidence to correlate the disorder with any particular aspect of growing conditions or husbandry. The literature on the effect of cultural factors on 'pepper spot' and related disorders has been reviewed by Cox (1977).

Causes

Pepper spot is a non-pathogenic disorder which cannot be associated with any fungus, bacterium or virus (Strandberg *et al* 1969; Geeson and Browne 1978; Walkey and Webb 1978) and is considered to be physiological in origin. Different varieties or selections of cabbage vary considerably in their susceptibility or tolerance to pepper spot and in each variety the incidence of symptoms varies between sites and between seasons (Strandberg *et al* 1969; Geeson and Browne 1978). In highly susceptible crops, some plants remain free of symptoms throughout storage. A list of winter white cabbage cultivars recommended for their relatively high tolerance to pepper spot has been compiled by the National Institute of Agricultural Botany (Anon 1980).

Black speck is increased by post-harvest fungicide dips (Kear and Symons 1973). However, Geeson and Browne (1979A) observed that although the early development of pepper spot was accelerated by some fungicide drench treatments, the incidence of symptoms had no effect on the extent of final trimming necessary for drenched and undrenched cabbages.

There have been reports from the USA (Isenberg *et al* 1971) and Germany (Bohling and Hansen 1977) that leaf necrosis disorders of cabbage can be prevented by controlled atmosphere storage. Geeson and Browne (1979B) confirmed that the incidence of pepper spot in several cultivars of white cabbage grown for storage in the United Kingdom could be reduced or eliminated by storing in atmospheres containing 5–6 per cent carbon dioxide, 3 per cent oxygen and the remainder nitrogen rather than in air.

References

Anon, (1980) List of cabbages 1980. Cambridge NIAB (Growers' Leaflet No. 2)

Betzeman, J and Commandeur, J C (1968) What is the best time to sow and plant out cabbages grown for storage? *Groent en Fruit*, **23**, p. 1577.

Bohling, H and Hansen, H (1977) Storage of white cabbage (*Brassica oleracea L.* var. *capitata*) in controlled atmospheres. *Acta Horticulturae*, **62**, pp. 49–50.

Cook, R J (1976) Blackspot in cabbage in Kent. *Plant Path.* **25**, pp. 181–6.

Cox, E B (1977) Pepper spot in white cabbage — a literature review. *ADAS Quarterly Review*, **25**, pp. 81–6.

Geeson, J D and Robinson, J E (1975) Damage will mean trouble in store. *Commerical Grower*, **4147**, pp. 1245–6.

Geeson, J D and Browne, K M (1978) Careful harvest is vital for white cabbage storage success. *Grower*, **89**(1), pp. 27–31.

Geeson, J D and Browne, K M (1979A) Controlled atmosphere keeps coleslaw crop greener. *Grower*, **92**(14), pp. 36–8.

Geeson, J D and Browne, K M (1979B) Effect of post-harvest fungicide drenches on stored winter white cabbage. *Plant Path.* **28**, pp. 161–8.

Isenberg, F M R, Oyer, E B and Engst, C B (1971) The effects of modified atmospheres plus physiologically active chemicals on cabbage storage life. *Acta Horticulturae*, **20**, pp. 7–18.

Kear, R W and Symons, J P (1973) Post-harvest deterioration of stored Dutch white cabbage. *Mededelingen Faculteit Landbouwetenschappen Rijunkiversiteit Gent*, **38**, pp. 1549–60.

Polegaev, Y I (1974) The nature of cabbage spotted necrosis. *Horticultural Abstracts* **44**, p. 3916.

Strandberg, J O, Darby, J F, Walker, J C and Williams, P H (1969) Black speck a non-parasitic disease of cabbage. *Phytopathology*, **59**, pp. 1879–83.

Theiler, R (1973) Blattpunktnekrosen bei Lagerkohl. *Gemuse*, **9**, pp. 326–30.

Walkey, D G A and Webb, M J W (1978) Internal necrosis in stored white cabbage caused by turnip mosaic virus. *Annals of Applied Biology* **89**, pp. 435–41.

Cabbage — large necrotic leaf spot (blackspot) (Plate 4.1)

After pepper spot, this is the most common necrosis problem of stored white cabbage in the United Kingdom. The large brown or black lesions may be 5–10 mm diameter and frequently coalesce to form irregular discoloured areas. Tissue in the centre of the spots becomes sunken and eventually collapses to leave a brown, papery membrane (Geeson and Browne 1978; Walkey and Webb 1978).

This disorder is the result of infection, usually early in the growing season, by aphid-borne turnip mosaic virus. Previously Glaeser (1970) isolated tulip mosaic virus from smaller leaf spots of cabbage in Austria and cauliflower mosaic virus has been reported to cause necrotic spots on white cabbage in Holland (Van Hoof, 1952) and in the United States of America (Natti, 1960).

References

Geeson, J D and Browne, K M E (1978) Careful harvest is vital for white cabbage storage success. *Grower*, **89**(1), pp. 27–31.

Glaeser, G (1970) Wodurch Enstehen Schwarze Stippen en Lagerkrut. *Der Pflanzenartz*, **23**, pp. 122–3.

Natti, J J (1960) Cabbage mosaic and leaf spotting in storage. *New York Agricultural Experimental Station Farm Research*, **24**, p. 13.

Van Hoof, H A (1952) Stip in Kool en Virusziekte. *Mededelingen van der Tuinbouw*, **15**, pp. 727–42.

Walkey, D G A and Webb, M J W (1978) Internal necrosis in stored white cabbage caused by turnip mosaic virus. *Annals of Applied Biology*, **89**, pp. 435–41.

Cabbage — vein streak (Plate 4.4)

This disorder of stored cabbage is similar to 'pepper spot' and appears as superficial brown or black markings on the epidermis along the midrib and petiole, occasionally spreading out along the larger veins (Geeson and Robinson, 1975), These symptoms are similar in many respects to those described in Holland as 'Koffiedik' (Jensma and Kraai, 1955) and 'Grijs' (Nieuwhof *et al* 1975). The cause of 'vein streak' is not known, but it is infrequent and rarely a serious problem.

References

Geeson, J D and Robinson, J E (1975) Damage will mean trouble in store. *Commercial Grower*, **4147**, pp. 1245–6.

Jensma, J R and Kraai, A(1955) Praktijkproeven met Witte Kool. *Mededelingen Instituut voor der Veredeling van Tuinbouwgewassen, Wageningen*, **61**, 35 pp.

Nieuwhof, M, Garretsen, F and Kraai, A (1974) Grey speck disease – a non-parasitic, post-harvest disorder of storage white cabbage (Brassica oleracea var Capitata L). *Euphytica*, **23**, pp. 1–10.

Cabbage — internal tipburn (Plate 4.5)

Although most types of cabbage can be affected, this disorder is most commonly seen on stored white cabbage. The margins of the inner heart leaves, especially around the vein endings, become papery and discoloured grey or brown (Nieuwhof, 1960; Geeson and Browne, 1978).

Susceptibility of headed cabbage to tipburn has been attributed to heredity factors (Dixon 1977; Nieuwhof 1960), although both Nieuwhof *et al* (1961) and Shafer and Sayre (1961) reported that the incidence of tipburn was correlated with high levels of nitrogen fertiliser and large head size.

References

Dixon, M H (1977) Inheritance of resistance to tipburn in cabbage. *Euphytica*, **27**, pp. 811–5.

Geeson, J I and Browne, K M (1978) Careful harvest is vital for white cabbage storage success. *Grower*, **89**(1), pp. 27–31.

Nieuwhof, M (1960) Internal tipburn of white cabbage. 1. Variety Trials. *Euphytica*, **9**, pp. 203–8.

Nieuwhof, M, Garrestsen, F and Wiering, D (1960) Internal tipburn in white cabbage. 2. The effect of some environmental factors. *Euphytica*, **9**, pp. 275–80.

Shafer, J and Sayre, C B (1961) Internal breakdown of cabbage related to nitrogen fertiliser and yield. *Proceedings of the American Society for Horticultural Science*, **47**, pp. 340–2.

Cauliflower — buttoning

Summary

Buttoning in cauliflowers is the premature exposure of the curd from the covering leaves. It occurs most commonly in early summer cauliflowers from autumn sowing. The Snowball and Danish types are susceptible to buttoning. The disorder is caused by a check in the growth of the plant after curd initiation. Its occurrence can be minimised by careful choice of sowing date and by planting good quality small plants.

Symptoms

The definition of buttoning is vague. Buttoning is only an extreme state in the range of curd conditions. A buttoned curd is a curd which is fully exposed before the age of development at which it is normally cut to eat. Exposure is due to restricted leaf growth. Skapski and Oyer (1964) found that in American cultivars a button has a low top weight, a low ratio of curd weight to surface area and low ratio of leaf weight to curd weight.

Occurrence

Buttoning is particularly important in early summer cauliflowers, and mainly in those sown in the autumn and over-wintered, although it can occur at other seasons when cultivars are grown to mature outside their normal curding period. It is influenced both by cultivar and environmental conditions. Aamlid (1952) found that cultivars with a low leaf number at maturity buttoned more readily but matured earlier than those with a high leaf number. The earliest cultivar studied had 36–40 leaves at harvest, the heaviest early cultivar 41–45 leaves and later cultivars up to 58 leaves.

Causes

The following effects of temperature have been noted at various stages of growth.

At seed germination.
Temperature differences between 1 and 20°C at seed germination have no effect on curd size or date of maturity (Parkinson 1952).

From germination to curd initiation

Temperature differences during the pre-curding stage of the plants affect both curd size and date of maturity, mainly because of the effect on stage of growth and leaf number at planting time. Low temperatures during the pot stage, $-1°C$ to $10°C$ which commonly occur in practice, give slower growth after planting and a longer period before curd initiation resulting in more leaves, fewer buttons and larger curds but slightly later maturity (Carew and Thompson 1948). Higher temperatures, e.g. long periods at around $10°C$ early in the life of the plant may bring it to the stage of curd initiation before planting out, and the plant will then button under the cooler conditions after planting.

After planting

Leaf area increase is less at low temperatures, giving smaller curd size and weight but slightly earlier maturity (Parkinson 1952). Carew and Thompson (1948) reported similar results with curd size, but maturity date was not affected by temperature. Aamlid (1952) found that alternating high day and low night temperatures produced a heavier curd than the same mean temperature evenly distributed.

Other factors which may have an influence on buttoning are:

Day Length

Day length appears to have little affect (Parkinson, 1952, Skapski and Oyer 1964). Aamlid (1952) found that short days (9 hours) during the seedling stage gave small plants at planting time, higher yields, less buttoning and later maturity than normal day length (12–15 hours), but the differences were not enough to have practical significance.

Nitrogen

Low nitrogen at the pot stage has generally given slower leaf production and a smaller plant at planting which reduced buttoning and, therefore, increased total yield but delayed maturity (Parkinson 1952, Sandwell 1961. Skapski and Oyer 1964). Delayed maturity with low nitrogen was also found by Aamlid (1952). In one of Salter's 1959 experiments, however, low nitrogen increased buttoning, possibly because the plants did not root out quickly after planting and were starved at that time. After planting, low nitrogen increased buttoning and reduced yield (Carew and Thompson 1948, Parkinson 1952). Maturity date was variably affected.

Watering

Watering treatments during the pot stage have had little effect. Carew and Thompson (1948) were able to reduce buttoning in one case only in four experiments by imposing a dry treatment at this stage. Parkinson (1952) produced no yield effect. Salter (1960) found a wet treatment in the pot stage better than a dry one. Adequate water is necessary after planting to prevent a check to growth.

Size of plant

Large plants button more readily but mature earlier than small plants because large plants have either initiated curds before planting or do so soon after. Although size does not necessarily indicate physiological age (Aamlid 1952), it is still a good guide under practical conditions. Jensma (1957) suggested small stem diameter of transplants as an indication that curd initiation had not taken place and Skapski and Oyer (1964) suggested planting plants of either less than 5 g in weight or with a stem diameter of 5 mm or less, or with less than 16 leaves as these were likely to be still in a pre-curd-initiation stage. If plants have initiated curds at planting time, they produce few additional leaves and small curds and any retardation causes buttoning (Aamlid 1952).

Trimming roots at pricking-out stage did not affect yield or buttoning (Parkinson 1952).

Buttoning in winter cauliflowers is a similar condition which occurs in a small proportion of the crop each year. There is a cultivar susceptibility.

References

Aamlid, K (1952) A study of cauliflower. Thesis, Univ of Maryland, (unpub).

Carew, J (1948) What makes cauliflower button. *Mkt. Grown. J.* **77**, pp. 12 and 39.

Carew, J and Thompson, H C (1948) A study of certain factors affecting buttoning of cauliflower. *Proc. Amer. Soc. Hort. Sci.* **51**, pp. 406–14.

Jensma, J R (1957) Teelt en veredeling van bloemkool. *Meded. Inst. Vered. Tuinb. Gewass.* **96**, pp. 1–61.

Nelder, J A (1955) The effect of size at planting out on subsequent development in summer cauliflower. Wellesbourne National Vegetable Research Station, *Research Communication,* **6**, p. 5.

Parkinson, A H (1952) Experiments on vegetative and reproductive growth in cauliflower. Wellesbourne Report of the National Vegetable Research Station for 1951. pp. 38–51.

Robbins, W R, Nightingale, C T and Schermerhorn, L G (1931) Premature heading of cauliflower as associated with the chemical composition of the plant. *N. J. Agric. Exp. Sta.* Bull. **509**.

Salter, P J (1960) The growth and development of early summer cauliflower in relation to environmental factors. *Hort. Sci.* **35**, p. 21.

Salter, P J (1959) The effect of different irrigation treatments on the growth and yield of early summer cauliflowers. *Hort. Sci.* **34**, pp. 23–31.

Salter, P J (1960) Effects of different soil moisture conditions during the seedling stage on the growth and yield of early summer cauliflowers. *Hort. Sci.* **35**, pp. 239–48.

Sandwell, I (1961) Propagation and planting of early summer cauliflowers, *Exp. Hort.* **4**, pp. 41–54.

Skapski, H and Oyer, E B (1964) The influence of pre-planting variables on the growth and development of cauliflower plants. *Proc. Amer. Soc. Hort. Sci.* **85**, pp. 374–85.

Walker, P (1953) Wind causes buttoning in cauliflowers. *Grower,* **39**, pp. 35–7.

Winter, E J (1952) Studies on cauliflowers and lettuce. Wellesbourne Report of the National Vegetable Research Station for 1951. pp. 31–7.

Winter cauliflower — scorch (Plates 4.6)

Summary

Heart leaf scorch occurs regularly in South West England in susceptible cultivars of Roscoff and winter hardy cauliflower. There is a cultivar susceptibility and the disorder may be associated with abnormal nitrogen metabolism.

Symptoms

The heart leaves initially develop normally but become stunted and brown at the tips. There is a reduction of curd size and secondary bacterial rotting may spread to the curd and sometimes cause death of the plant. Symptoms develop two to three months after planting and reach their maximum at time of heading. Mild humid conditions accentuate the disorder. Plants with scorch are more liable to frost damage.

Occurrence

This disorder has been known for some years and is found most commonly in South West England.

One or two cases occur every year. Within affected crops it can be serious as up to 25 per cent of plants may be damaged.

Causes

Susceptibility to scorch appears to be an inherited characteristic and the extent to which it affects crops varies with season. The extent to which any cultivar develops scorch is partly related to high nitrogen use in the base dressing. Inorganic nitrogen fertiliser appears to be worse in causing scorch than organic nitrogen, possibly by its quicker action in raising the nitrogen level.

Applying molybdenum has been shown in trials to reduce scorch indicating a connection with nitrogen metabolism in the plant.

A similar condition but accompanied by translucent curds has been reported in Germany and attributed to calcium deficiency. Here experiments showed an increasing incidence as relative humidity increased and calcium uptake decreased (see curd discolouration in cauliflowers page 43). A similar calcium imbalance may be associated with scorch.

References

Jenkinson, J G and Williams, Audrey M (1967) Stem rot and scorch in winter cauliflower. *Expl. Hort.* **17**, pp. 57–66.
Wiebe, H J and Krug, H (1973) Calcium deficiency in cauliflowers, symptoms and causes of damage. *Gemuse,* **9**(5), pp. 140–2.

Winter cauliflower — stem rot (hollow stem) (Plate 4.7)

Summary

Hollow stem occurs widely at a low level in some cultivars. It is promoted by vigorous early growth and this is influenced by the use of large amounts of nitrogen fertiliser.

Symptoms

A brownish area or hollow space occurs in the centre of the stem resulting from the collapse of the internal parenchyma tissue. Secondary invasion of the damaged tissues by bacteria may lead to extensive rotting which can extend up into the curd, and can also weaken the stem sufficiently to cause the plant to collapse.

Occurrence

This disorder is widespread and has been known in brassicae for many years. It occurs widely at a low level and only occasionally is it serious. Within a particular cultivar it is more likely in early maturing plants.

Causes

Stem rot is an inherited factor and incidence varies appreciably in different cultivars. The basic cause is the failure of the internal tissues to keep up with the increase in stem diameter. Any factor which increases the rate of growth is liable to increase the incidence of stem rot. Thus large applications of nitrogen fertiliser, particularly when applied early, increase stem rot. Plants which mature early and therefore have grown rapidly are more susceptible.

The incidence of stem rot in commercial crops can be minimised by using less susceptible cultivars and moderate applications of inorganic nitrogen.

A similar disorder occurs in sprouting broccoli in the USA and Canada where it was observed that increasing the nitrogen nutrition, earliness of maturity and wider spacing increased the incidence.

References

Jenkinson, J G and Williams, Audrey M (1967) Stem rot and scorch in winter cauliflowers. *Expl. Hort.* **17**, pp. 57–66.

Hipp, B W (1974) Influence of nitrogen and maturity rate on hollow stem in broccoli. *Hort. Sci.* **9**, pp. 68–9.

Cutliffe, J A (1972) Effects of plant spacing and nitrogen on incidence of hollow stem in broccoli. *Can. J. Plant Sci.* **52**, pp. 833–4.

Cauliflower — localised curd discoloration

Summary

There is reduction in curd size and brownish water-soaked areas appear on the surface of the curd. The brown areas extend down the inside of the branches of the flower into the upper part of the main stem. Commonly these symptoms are associated with stunting and distortion of the leaves immediately surrounding the curd.

Causes

The disorder is frequently linked with boron deficiency but the development of translucent areas of curds and marginal necrosis of the young leaves has been reported to be caused by calcium deficiency (Wiebe and Krug 1973). The latter condition occurred under high humidity with reduced foliar transpiration and was aggravated by high nitrogen fertiliser use and high conductivity. Under controlled experimental conditions a low relative humidity (40 per cent) prevented the development of the symptoms.

Mildew can cause a similar discoloration to flower stalks. The occurrence of the disorder has been linked·with water soaking or frost damage to the curd which is quickly followed by an invasion of soft rotting bacteria.

References

Anon, (1969) Diseases of vegetables: boron deficiency. London, MAFF Reference Book 123, p. 26.

Wiebe, H J and Krug, H (1973) *Gemuse,* **9**(5), pp. 138 & 140–2.

Spinach — tipburn

Brown marginal necrosis appears on the central unexpanded leaves. The necrotic zone is usually only a few millimetres wide, but the adjacent tissue may be chlorotic.

The symptoms usually appear shortly before harvesting, and the distribution in the field is very erratic. Symptoms have been seen regardless of the time of harvest from spring to autumn. This disorder is more common in Essex than in other spinach-growing areas and occurs much more frequently on some holdings than others. A limited survey in Essex in 1973 did not identify any common factors except for a possible association with water stress.

A somewhat similar condition, usually distinguishable, can result following infection with cucumber mosaic virus and is known as 'blight'.

Root crops and brassicae

Root crops and brassicae — strangles (Plate 5.1)

Summary

A constriction of the hypocotyl near soil level caused by flexing and twisting of the young plant during strong winds or drying out after singling.

Symptoms

Strangles appears in young plants, in sugar beet, red beet, mangels and similar crops — it usually occurs soon after singling. The hypocotyl becomes constricted at or just above soil level. The parts of the root above and below the constriction continue to grow slowly. The effect of the constriction is to stunt the growth of the plant. In root crops the upper part may eventually break off.

Occurrence

Plants showing strangles are fairly common, particularly in western districts, but the proportion of plants affected in any one crop is usually low. Occasionally 20 per cent of a field may be affected. The condition is commonest at seedling and singling stages of growth in fields exposed to strong winds and when the surface soil is dry and hard.

Causes

The damage occurs some time before the symptoms are visible. Flexing and twisting of the young plants by strong winds is the commonest cause. 'Damping-off' fungi (Pythium and Rhizoctonia spp) may also contribute. Another cause, quite common when singling was the usual technique was the drying out of the cortex of the newly exposed hypocotyl. Herbicides may cause similar symptoms but are usually found lower on the tap root.

Importance

If the affected plants are randomly distributed over the crop and the number is not greater than 10 per cent of the total then the root yield of sugar beet or mangels is unlkely to be seriously affected. In conditions where strangles are likely to occur, replacing soil around the young plants, when this is possible, will reduce the incidence.

References

Anon, (1969) Diseases of vegetables. London, MAFF. Reference Book 123.
Anon, (1979) Strangles of sugar beet and mangels. MAFF Leaflet 547.
Boyd, A E W (1966) Sugar beet strangles. *Edin. Sch. Agric. Tech. Bull.* 26.

Fruit Crops

Apple — Cox Disease (Plate 6.1)

Summary

Cox's Orange Pippin apple trees on M9 or M26 root stocks can be affected by Cox Disease during their non cropping years resulting in areas of bare wood on extension shoots. Partial ring barking can alleviate the symptoms. Once the trees commence cropping the severity of the disorder decreases. The syndrome can appear in older trees carrying a light crop and sometimes following harvest.

Symptoms

In July and August symptoms first appear at the tips of strongly growing shoots which become chlorotic and the young leaves up rolled. Later in August and September leaves begin to show purple colouring starting with the older leaves, followed by necrotic blotching and leaf fall. The early loss of leaf leads to bud abscission in the early spring and bare areas on the shoot.

Occurrence

Symptoms of the Cox Disease syndrome were reported during the 1960s from young plantations of Cox's Orange Pippin growing on the dwarfing root stocks M9 or M26. Investigational work by the Dutch was reported by Oud (1968) and Visser (1968). A scheme relating the various symptoms associated with the Cox Disease syndrome was proposed by White (1972).

The disorder is linked closely with the new plantations of Cox trees on specific root stocks (M9 and M26) and severe symptoms are usually observed on strong growing young trees. Within a block of young Cox trees the occurrence of the symptoms is random and unpredictable. If a tree has shown spring symptoms it is likely that it had summer symptoms the year before. Once young trees commence cropping there is apparent recovery from the disorder, although White (1972) stated that during a light cropping year summer symptoms of Cox Disease can reappear in more mature trees. The areas of bare wood within the tree can present problems in preventing development of a suitable productive framework for the tree.

Causes

Investigations in the Netherlands by Visser and Oud suggested that there was root death in the peripheral root system of the trees affected by Cox Disease and this was associated with poor carbohydrate translocation and consequential imbalance. The accumulated carbohy-

46

drates in the leaf tissue caused the summer leaf purpling symptoms. The mineral composition of leaves from affected trees was found by Dutch, East Malling and ADAS investigators to be lower than the levels in normal trees when expressed as a percentage of the dry weight. The actual weights of nitrogen, calcium and magnesium in Cox leaves showing disease were respectively 34, 48, 38 per cent less than in normal leaves (White, 1970). It was concluded that there was reduced rate of transfer of nutrients to the yellowing spur leaves. Hansen (1974) reported that in sand culture, trees with a high boron supply suffered from an increased incidence of Cox Disease, with delayed flowering, but abscission and leaf purpling in late seasons.

Veen and Locher (1972) attempted to examine root death relative to carbohydrate movement in young trees on M9 grown in sand and water culture.

There was no irregularity in the transport of carbohydrates across the graft union of Cox on M9 or M26. A depleted supply of carbohydrate, induced by shading, did not cause root death in these experiments. These workers tentatively concluded that Cox Disease was connectd with a low auxin level in the root system which also motivates photosynthate transport into the roots.

Partial ring barking alleviated summer symptoms in the Dutch experiments (Oud 1968). White (1972) found that half-ringing above the graft produced larger and darker leaves with better quality blossom and with reduced summer symptoms whilst ringing below the graft aggravated the symptoms.

Investigational work by ADAS in 1970 and 1971 on four and five year old Cox trees confirmed that ring barking in two consecutive years reduced foliar symptoms in August and September (Annual Report, Luddington EHS 1971). As this treatment had little effect on the mineral composition of the leaves or xylem sap it was concluded that it was not a nutritional problem. Affected leaves on 'diseased' trees had a higher dry matter composition with a total soluble carbohydrate level 12 per cent above that of normal leaves. The rate of leaf respiration was unaltered by treatment or by occurrence of symptoms, whereas there was a higher rate of photosynthesis in the leaves of the ring-barked trees, and affected leaves showed considerably reduced rates of photosynthesis. Using biochemical analysis and isotope techniques it has been shown that the effect of ring-barking on Cox Disease and the translocation of metabolites in deblossomed trees has given freer movement of ^{32}p introduced into the trunk of ringed trees. More soluble carbohydrates were recovered from the root system and the amino acid activity in June and July was enhanced by ringing. Cox Disease was not prevalent on these trees during the two seasons of extensive biochemical work, but the study showed the stimulatory effect of wounding or ringing on metabolite movement between root and shoot.

The critical inter-relationship between the metabolic activity of root and shoot relative to the Cox Disease syndrome was indicated by the observations of White (1972). Cox trees closely planted at 30 cm showed less of the disorder compared with trees spaced at 240 cm. This effect is a consequence of increased tree to tree competition and reduced vigour.

References

Anon. (1971) Control of Cox disease, 1969–71 Progress Report. *Ann. Rept. Luddington EHS.* pp. 18–25.

Hansen, P (1974) The effect of boron upon leaf development and growth of apple cultivar Cox's Orange Pippin. *J. Hort. Sci.* **9**, pp. 211–6.

Cud, P (1968) New discoveries made on Cox's diseases. *Fruitteelt,* **58**, pp. 1172–4 and pp. 1194–6.

Veen, B W and Locher, J Th, (1972) Some investigations on Cox's disease in Cox's Orange Pippin apple trees, grafted on dwarfing rootstock M IX. *Neth. J. agric. Sci.* **20**, pp. 285–300.

Visser, J (1968) Investigation of the influence of soil type on the occurrence of Cox disease in the Ijsselmeer-polders. *Flevoberichen (Zwolle),* **56**.

Visser, J, Locker J Th and Brouwer, R (1971) Effects of aeration and mineral supply on growth and mineral content of shoots and roots of apple trees. *Neth. J. agric. Sci.* **19**, pp. 125–37.

White, G C (1969) Cox disease: a foliage disorder of apple trees. *Ann. Rept. E Malling Res. Sta.* pp. 187–8.

White, G C (1972) Cox disease. *Ann. Rept. E Malling Res. Sta.* pp. 129–31.

Apple — leaf spot and leaf drop (Plate 6.2)

Summary

The main factors affecting leaf spot and leaf drop in Golden Delicious apples are temperature, light intensity and leaf age. Temperatures above 21°C at low light intensities are particularly harmful. Older leaves are affected first. An imbalance in mineral composition has been noted.

Symptoms

Older leaves on apparently healthy trees suddenly show irregularly-shaped, but well-defined brown necrotic spots, varying in diameter from 1 mm to more than 29 mm. Generally, the appearance of 'spots' is followed by leaf abscission and leaf drop, but not in all cases. In severe cases of drop, trees can lose more than 75 per cent of their leaves with only young ones left near the apex of the shoot, with a consequential effect on cropping in the following year.

Occurrence

The disorder was widely reported on the Continent in the late 1960's, affecting mainly the cultivar Golden Delicious. Similarly, leaf spot symptoms were observed in Cox's Orange Pippin (Cox Spot and Frog Eye) and in Winston (Clijsters, 1970). A definite relationship with rootstock type has been established, although trees with vigorous growth are more affected by leaf drop than weaker trees. Observations suggest that the problem is aggravated by root death.

Causes

The intensity of the disorder differs from year to year and is probably associated with climatic factors (Jonkers, 1973). Dry, hot weather in July and August is generally regarded as a main factor if combined with sudden changes of temperature. Jonkers (1969, 1963) has shown that a temperature of 21°C is optimal for promotion of the disorder, especially under conditions of low light intensity. Variation in the amount of soil water does not affect leaf spot. Deficiency levels of magnesium, manganese and calcium and an excessive level of potassium have been found in spotted leaves, but foliar application of mineral solutions has not prevented the appearance of the symptoms. However, carbamate fungicidal sprays containing manganese have been found to alleviate the symptoms.

More recent work by Kender and Jonkers (1975) has indicated the involvement of gibberellins in the promotion of leaf spot.

References

Clijesters, H (1973). Leaf fall in Golden Delicious. *Fruitleeltbled,* **14**, pp. 11–13 (Hort. Abst. (1971) 5660).

Jonkers, H (1969) Leaf spot and leaf drop in Golden Delicious on M1X. *Fruitteelt,* **59**(33), pp. 1014–7.

Jonkers, H (1973A) Leaf spot and leaf drop in the apple cultivar Golden Delicious: a physiological disorder. *Neth. J. agric. Sci.* **21**, pp. 171–80.

Jonkers, H (1973B) Review of leaf spot and leaf drop: a physiological disorder of Golden Delicious apples. *Scientia Hort.* **1**, pp. 231–7.

Kender, W J and Jonkers, H (1975) Gibberellin promotion of physiological leaf drop in Golden Delicious apple leaves. *Neth. J. agric. Sci.* **23**, pp. 126–30.

Apple Fruit Disorders

The literature on apple fruit disorders is wide ranging and has been extensively reviewed by Carne (1948), Faust *et al* (1969), Wilkinson (1971) and Fidler *et al* (1973). Consequently only a brief description of each disorder is given together with a brief note on the causal factors. The disorders can be classified as follows:

Pre-harvest disorders

Bitter pit
Cracking and russetting
Crinkle
Watercore

Storage disorders

Carbon dioxide injury
Coreflush
Freezing injury
Low temperature injury   a) internal breakdown
                            b) ribbon scald

Oxygen deficiency injury
Senescent breakdown
Superficial scald

Pre-harvest disorders
Apple — bitter pit (Plate 6.3)

Symptoms

Bitter pit is a condition in which the flesh of the apple is disfigured by small brown corky-looking dry areas up to 5 mm in diameter. These areas which may appear as depressions in the skin or as localised areas in the outer cortex of the apple just below the skin occur most frequently at the calyx end of the fruit. The abnormal cells in the 'pits' are collapsed, but the cell walls are normal and there is no true cork formation. These cells retain starch long after starch has disappeared from the healthy cortical cells. Although the disorder can occur in fruit on the tree, it usually appears only after a period in store.

Causes

An imbalance in the mineral composition of the fruit is the suspected cause of bitter pit in susceptible cultivars. A low calcium content with high potassium (and magnesium) is frequently found in fruit from affected samples. Light soils and irregular water supply influence the availability and mobility of calcium and boron. Vigorous vegetative growth can compete for the limited supply of calcium. Routine spraying with calcium salts during fruit development is effective in alleviating the disorder. In some cultivars fruit calcium content is further supplemented by post-harvest immersion in calcium chloride solutions. Although boron deficiency is relatively uncommon in the United Kingdom, occasional responses to solubor sprays have been noted. However, this treatment should be used with caution since it may accelerate fruit ripening.

Apple — cracking and russetting (Plate 6.4)

Symptoms

Depending on cultivar and seasonal severity, the symptom expression can vary between a typical russetting of the skin to distinct cracking with associated cork formation.

Causes

Cool weather or water stress early in the growing season is particularly conducive to russetting, whilst rapid expansion following a rainy period can lead to cracking. Certain pesticide sprays can also increase russet development. Recently poor mobility of boron in the fruit has been implicated and although repeated foliar application of boron (as solubor) has improved skin finish in Egremont and Cox in some situations, there are many other causes of this type of blemish apparently unrelated to nutritional factors. Encouraging results have been achieved in experimental work using synthetic growth regulator sprays (GA 4 and 7) to reduce skin russetting.

Apple — crinkle (Discovery) (Plate 6.5)

Symptoms

Dark, slight sunken areas on the coloured part of the fruit surface. There is an associated breakdown of tissue beneath the surface injury, but this does not penetrate very deeply. Usually only a proportion of the fruits are affected and these are almost invariably the most highly coloured. Symptoms may not be reported until just before or during picking, but may well be present earlier.

Causes

The disorder was seen on one Worcestershire farm in 1977 and 1978. In both years fruit calcium levels were found to be low and this may be a contributory factor. The disorder was also seen in 1977 in Suffolk. The symptoms bore some similarity to crinkle, a disorder described for Australian apples by Carne (1948). An examination in 1977 of several other Discovery orchards in Worcestershire and Warwickshire did not reveal any trees with the disorder and this seemed to rule out weather conditions as being directly involved.

Apple — watercore (Plate 6.6)

Symptoms

In the early stages symptoms appear as translucent (water-injected) tissue in the vicinity of vascular bundles or near carpel walls. The intercellular spaces are filled with sap. In moderate to severely affected fruit internal breakdown develops in the areas of the water core. During storage watercore can disappear from slightly affected apples, but in more severely affected fruit it may be succeeded by excessive internal breakdown.

Causes

Watercore tissue has been found to contain some fermentation products. The intercellular spaces are filled with sorbitol which accumulates because the tissue does not have the capacity to convert it to fructose. Sorbitol is the major metabolite of photosynthesis in apple leaves at the time of watercore development. Cultural practices which encourage a high leaf to fruit ratio and consequently increase the potential supply of assimilates to the fruit increase the incidence of watercore. The disorder is often associated with low concentrations of calcium in fruit. This may lead to an increase in cell permeability, thereby leading to the outward leakage of cell sap into the intercellular spaces.

Storage disorders
Apple — carbon dioxide injury (Plates 6.7 and 6.8)

Symptoms

The external symptoms are well defined, sunken areas of the skin which are initially dark green, but turn brown and finally almost black.

The disorder is first seen internally as pockets of dark-brown discoloured tissue often associated with the vascular tissues in the cortex of the fruit. The damaged tissues are initially moist and brown, but cells collapse and dry out, creating brown cavities within the cortex. The injured tissues have a characteristic spongy feel.

Causes

The damage is caused by excessively high carbon dioxide conditions during storage. It is occasionally seen on Bramleys and is aggravated by low temperatures and poor air movement.

Apple — coreflush (Plate 6.9)

Symptoms

This disorder is frequently seen in refrigerated air and controlled atmosphere storage as a pinkish-brown discoloration of the core tissue between the carpels and loculi. In severe cases cavities will form in the injured tissue.

Causes

This disorder is promoted in most cultivars by a range of conditions in store (e.g. low temperatures, relatively high carbon dioxide with low oxygen conditions, and extended storage). Cool, sunless growing conditions, premature harvesting, low leaf nitrogen and high fruit and leaf potassium have been associated with the occurrence of coreflush. Cox's Orange Pippin apples are more susceptible to coreflush following applications of daminozide in the orchard.

Apple — freezing injury (Plate 6.10)

Symptoms

The affected fruit appears translucent and on thawing the internal tissues become water-soaked. The vascular strands become dark brown. Cone-shaped sectors of damaged tissue spread out from the core.

Causes

Freezing injury is usually caused by a failure to calibrate thermometers and thermostats correctly. Alternatively, irregular stacking can cause local cold spots in part of the store. For pear storage it is essential to fit a second overriding thermostat in case the main thermostat should fail. Fruit which has been held for only a few days at temperatures a few degrees below its freezing point may recover completely on thawing. The risk of freezing is increased in pears of low soluble solids content.

Apple — low temperature injury: internal breakdown (Plate 6.11)

Symptoms

Low temperature injury usually appears as a browning in the cortical region with streaks of darker brown in the vascular tissues. The boundary between affected and sound tissue is diffuse and there is usually a zone of 2–3 mm of clear tissue immediately below the skin. The tissue remains moist and does not become dry and mealy as in some other forms of breakdown. As the disorder progresses, the skin eventually becomes discoloured and apparently waterlogged, giving a translucent appearance.

Causes

This disorder is induced by prolonged exposure to low temperatures which are below the minimum that is recommended for the cultivar, but are above the freezing point of the tissue. Its development is preceded by the accumulation of oxaloacetic acid which is thought to block the functioning of the tricarboxylic acid cycle.

Low phosphorous and low calcium levels make the fruit more likely to be affected by low temperature breakdown and some measure of control of the disorder has been obtained by applications of orchard sprays containing calcium chloride and calcium phosphate.

Apple — low temperature injury: ribbon scald (Plate 6.12)

Symptoms

The disorder occurs on a few cultivars as a ribbon-shaped brown band on the skin with well-defined margins, extending 2–3 mm into the cortex.

Causes

A form of low temperature injury.

Apple — oxygen deficiency injury (Plate 6.13)

Symptoms

This disorder is characterised by an alcoholic odour and the taste of the fruit is tainted. In severe cases a brown discoloration of the cortical tissues of the fruit occurs and the surface may become indented.

Causes

This disorder is usually caused by failure to maintain recommended minimum oxygen levels in controlled atmosphere stores. Such conditions arise if the gas sampling lines are leaking or the oxygen meter is faulty. Independent regular cross-checks of oxygen concentration are recommended. The accumulation of alcohol in the fruit is induced by anaerobic fermentation. Providing a critical alcohol level has not been maintained for too long, ventilation and an increase in fruit temperature will disperse the alcohol and the fruit may recover. However, apples which contain over 120 mg/100 g fresh weight are likely to become senescent before the alcohol falls to an acceptable concentration.

Apple — senescent breakdown (Jonathan or mealy breakdown) (Plate 6.14)

Symptoms

Skin becomes dull brown, cortical tissue is often off-white to yellow and mealy in texture, giving a spongy feel to the whole fruit. The outer cortical tissue becomes brown and under high humidity storage conditions splitting may occur. In Cox and other varieties the disorder tends to orginate at the calyx end of the fruit. It is associated with low calcium and phosphorus concentrations in the fruit.

Causes

At present, the accumulation of acetaldehyde is regarded as being responsible for initiating the breakdown of tissue, resulting in browning. The disorder occurs in over-mature apples due to late picking and extended storage and can be controlled by the application of calcium sprays, picking at a slightly immature stage and adoption of controlled atmosphere rather than air storage.

Apple — superficial scald (Plate 6.15)

Symptoms

Symptoms occur as fruit ages in store. At first these appear as faintly bronzed areas of skin in which the lenticils stand out prominently against small 'islands' of healthy epidermal cells. The disorder progresses and becomes darker until extensive patches of the skin are affected, which may become sunken. The form of the disorder varies with different cultivars. The early stage described above is sometimes termed 'rugose' scald, whilst the more general injury around the stalk is called 'stem end browning'. Scald is occasionally recorded on pears and certain red-skinned apples, but is most common on green-skinned apples such as Bramley's Seedling, Edward VII and Granny Smith.

Causes

Superficial scald is thought to be induced by high evaporation conditions and is most severe after dry, sunny summers. It is increased by early picking, delayed store cooling, poor ventilation in the store and the use of bulk bins rather than boxes to hold the fruit. The localised accumulation of the volatile oxidation products of alpha-farnesene, a constituent of the apple cuticle, is believed to be the primary cause of the disorder. The use of post-harvest drenches or dips containing the anti-oxidants, ethoxyquin or diphenylamine provides a very effective control of superficial scald in stored apples.

Another form of scald associated with over-storage in air is sometimes seen on Idared, Golden Delicious, Laxton's Superb and other varieties. This is described as senescent scald and does not always respond to post-harvest treatment with anti-oxidants.

References

Carne, W M (1948) The non-parasitic disorders of apple fruit in Australia. *Bull. CSIRO Aust.* p. 238.

Faust, M, Shear, C B and Williams, M X (1969) Disorders of carbohydrate metabolism of apples. *Bot. Rev.* **35**, pp. 169–94.

Wilkinson, B G (1970) Physiological Disorders of fruit after harvesting. *In:* A C Hulme (Ed.) The Biochemistry of fruits and their products Vol. 1, London and New York, Academic Press.

Fidler, J C, Wilkinson, B C and Sharples, R O (1973) The biology of apple and pear storage. Farnham, Commonwealth Agricultural Bureau, England.

Strawberry — tipburn

Summary

Tipburn in young leaves of strawberries, although widespread in some cultivars grown outdoors in summer, is seldom a serious problem, but it can be severe in rapidly growing plants in glasshouses. The cause is an inadequate rate of transport of calcium.

Symptoms

Grey, water-soaked areas appear on the tips of emerging leaves, developing into tip or marginal necrosis. During full leaf expansion, undamaged areas of the leaf become distorted. Flowers may also be affected.

Occurrence

This disorder was first described in 1933, and in early reports of its occurrence in plants grown in nutrient solution it was attributed to boron or calcium deficiency. Tipburn often appears in rapidly-growing plants in glasshouses, but although often widespread is seldom severe outdoors. Susceptibility differs between cultivars, Grandee and Gourmella being extremely sensitive, Cambridge Favourite intermediate in sensitivity, and Gorella much less sensitive. Probably no cultivars are immune if conditions are sufficiently extreme.

Causes

Experiments at Long Ashton show that tipburn is caused by localised shortage of calcium in emerging leaves (Mason and Gutteridge, 1974). It can be induced by deficiency of calcium in the rooting medium, but the usual cause is an inadequate rate of transport of calcium into unemerged leaves because of (a) an excessive concentration of salts in the rooting medium or (b) excessively low relative humidity at night-time, both factors resulting in reduced root pressure and thus a reduced flow of water into the bud. When salt concentrations are excessive, high relative humidity in the daytime may aggravate any existing shortage of calcium in newly emerged leaves and worsen tipburn by checking transpiration and thus making calcium transport dependent on root pressure flow in the daytime, as well as at night. Tipburn is alleviated by low salt concentrations and high night-time, combined with low daytime, humidities (Bradfield and Gutteridge, 1973).

References

Mason, G F and Gutteridge, C G (1974). The role of calcium, boron and some divalent cations in leaf tipburn of strawberry. *Scientia Horticulturae*, **2**, pp. 299–308.

Bradfield, E G and Gutteridge, C G (1979). The dependence of calcium transport and leaf tipburn in strawberry on relative humidity and nutrient solution concentration. *Ann. Bot.* **43**(3) pp. 363–72.

Salad Crops

Lettuce — tipburn (Plates 7.1 and 7.2)

Summary

At least four types of symptom expression are described for the tipburn complex in lettuce. At certain seasons and under certain climatic conditions the disorder can be severe with considerable losses in outdoor and glasshouse crops. Cultivars differ in their susceptibility to the condition. Moisture relations, conductivity, uptake and mobility of calcium and rate of leaf development can be involved in the disorder.

The disorder can be partially prevented by control of water loss from the plants and by supplying additional calcium to the susceptible leaves (although this may not be commercially viable).

Symptoms

In general, tipburn can be defined as the collapse of epidermal cells mainly at the leaf margins with subsequent development of brown necrotic tissue.

In the Dutch literature van der Hoevon (1965) classified four types of tipburn; the separation is not universally used and depends on the conditions causing the disorder.

Occurrence

The disorder was widely reported in the early 1950s in both the Netherlands and the USA. The importance of the moisture status of plants and of root activity were discussed by van den Kloes (1952), Grainger (1952), Western (1953) and van den Ende (1954). The symptoms were considered to be associated with unbalanced growth of the lettuce plants. Tipburn has been associated with and is considered to be induced by similar conditions to other disorders of lettuce described as russet, brown blight, rib blight, red heart, brown rib and internal breakdown. Spotted wilt is a similar disorder induced by infection with spotted wilt virus.

There are several reports of differing cultivar susceptibilities (Stewart and Foster, 1975, Jenkins, 1959, van de Hoeven *et al*, 1967, and Corgan and Nakayana, 1967).

The four types of tipburn are summarised below:

Normal tipburn

This occurs mainly in the spring during a period of rapid growth when transpiration by the crop exceeds water uptake especially under sunny conditions. It occurs mainly on the leaves enclosing and immediately beneath the head. The leaf margin becomes flaccid and has a water soaked transparent appearance. As water is evaporated from this area the margins turn brown and finally become dry and papery, sometimes transparent. Secondary rotting may occur.

Causes

Crops grown at high night temperatures (10°C) and in soils with a high conductivity are more susceptible to this form of tipburn. The incidence can be reduced by shading when there are sunny conditions and maintaining a high soil moisture content.

Dry tipburn (Marginal tipburn)

This problem occurs in both protected and outdoor crops. The symptoms occur in the older leaves of the plant particularly in the autumn and winter. The problem is more severe in crops which have reached maturity slowly. The disorder is characterised by small brown necrotic lesions along the leaf margins which subsequently become curled and wavy.

Causes

Dry tipburn occurs when water loss by transpiration exceeds water uptake and is promoted by sudden checks in growth such as exposure to low temperature. The problem can be alleviated by encouraging good water uptake, the production of active roots and good growing conditions. Reduction in the loss of water from the leaf by shading or lowering temperature also reduces the incidence by lowering transpiration losses. It can occur at any time of the year in drying conditions, i.e., strong sunlight and wind. It often appears in crops which have reached maturity. In outdoor crops, soil preparation should encourage moisture retention and extensive root development. In adverse conditions, mature crops should be harvested as soon as possible.

Dry tipburn frequently occurs in protected crops in early spring when bright sunshine and ventilation in a drying wind impose an impossible load on an inactive root system in cold soil. It can be avoided by 'flash damping' prior to careful ventilation. In general, good soil conditions should be maintained with attention to pH, salt concentration and moisture content.

Veinal tipburn (Glassiness)

This disorder occurs only in protected crops mainly in autumn and winter. (November to early January). The symptoms can occur in the younger leaves inside the head so that the symptoms are not apparent until the head is cut. Glassiness symptoms often show as well-defined areas, between the veins, near the margins of the outer leaves of younger plants, when the environmental conditions are not too severe. In more severe conditions, the centre leaves of young plants may be more generally water-soaked. As the lettuce matures, the centre leaves forming the heart are generally affected. If the waterlogging is of short duration (a few hours) cells may return to normal, but if of long duration the affected area may become necrotic.

Causes

Veinal tipburn is caused by excessive water uptake in relation to the transpiration loss by the plant. Glassiness is stimulated by factors which inhibit transpiration (such as low leaf temperature and high relative humidity) and those which promote water uptake (moist, warm soil and low salt concentration and extensive root system).

Glassiness can often be avoided by controlling relative humidity, e.g., a period of ventilation at the end of the day will exchange air loaded with moisture of transpiration for ambient air which will usually be drier. Adequate control of relative humidity may be difficult without the expense of heating. Irrigation and high volume spraying should be restricted, if possible, to periods of bright, drying weather. Particular care is needed where direct fired paraffin heaters are used and in plastic film structures. Some varieties are more susceptible than others, e.g., Pallas and Nordine. Adequate ventilation, especially at night when the temperature does not fall sufficiently to bring heat on, is an essential part of cultural avoidance of glassiness.

When glassiness is apparent at dawn, immediate action is needed to increase transpiration. The air temperature should be raised by at least 5°C for up to two hours before ventilating freely. These heating/ventilating cycles can be repeated if necessary, but are expensive.

Speckled (Latex tipburn)

This problem normally occurs with some late spring varieties manifesting symptoms in the spring in the young leaves enclosed inside the head. Under certain conditions the latex ducts (laticifer cells) in the leaf become ruptured so that the latex is excreted onto the leaf surface. The globules become oxidised on exposure to air after which the underlying and surrounding cells die. The latex exudation normally occurs in the morning. This effect is well described by Tibbits *et al* (1965) and Olson *et al* (1967 and 1969).

Causes

It appears that under some conditions the disorder is more likely to occur as plants enter their reproductive stage (van der Hoeven 1965) though there are conflicting views on this. Large day/night differential temperatures, high humidity and adequate soil water lead to conditions causing guttation of latex as a result of high water intake through the roots and low transpiration from the leaves. Harvesting as soon as the crop is mature avoids the problem.

Experimental work on tipburn

The comparison of different radiation regimes has shown that a relationship exists between rapid leaf development (initiation and lateral expansion) promoted by long day regimes and the occurrence of tipburn (Olson 1968). Tibbitts and Rao (1968) showed that leaf injury was most severe in long days and high light intensity. Leaf initiation had to exceed 1.5 leaves per day before plants became susceptible. The width/length ratio for leaves developing injury was always less than one. These findings contribute to the conclusion that tipburn results from an abnormal laticifer development and subsequent rupture in rapidly expanding leaves (Olson *et al* 1969).

Analysis of lettuce showed that leaves with tipburn contained less calcium (Leh 1970) and more free amino acids than normal leaves. The disorder was associated with an accumulation of aspartic acid, glutamic acid and their amides. It was induced by high nitrate and low calcium (Ashkar and Ries 1971). Thibodeau and Minotti (1967) reported that foliar sprays of calcium nitrate or calcium chloride controlled tipburn when applied to susceptible immature leaves. Corgan and Cotter (1971) tested the effect of 14 chemical treatments on tipburn. Five chemicals including calcium chloride and daminozide (B-Nine, Alar) reduced the symptoms. Results on the effectiveness of soil or foliar application of calcium differ. Kruger (1966) and Corgan and Cotter (1971) found calcium sprays effective whereas Borkowski and Ostrzycka (1973) found a soil treatment with 3 per cent Lena (a fertiliser containing Ca, K, and Mg) better than spraying with 0.5 per cent calcium chloride or calcium nitrate or the addition of 2 g/l calcium carbonate to the soil. Calcium is known to be relatively immobile within the plant and therefore the effectiveness of any treatment is determined by calcium reaching the sites of action in sufficient quantity. This becomes difficult after head formation using foliar sprays of calcium.

Work suggests that high localised auxin activities predispose the plant to tipburn development (Crisp, *et al* 1976) possibly by creating a demand for calcium which the plant cannot supply. This enhanced auxin activity may arise because of high concentrations of naturally occurring polyphenols in the plant and which interfere with auxin oxidase activity. Evidence has been obtained that those lettuce cultivars more resistant to tipburn contain lower concentrations of polyphenol chlorogenic acid (Collier, *et al* 1979).

References

Ashkar, S A and Ries, S K (1971) Lettuce tipburn as related to nutrient imbalance and nitrogen composition. *J. Amer. Soc. Hort. Sci.* **96**(4), pp. 448–52.

Borkowski, J and Ostrzyoka, J (1973) The control of blossom-end rot of tomatoes and tipburn in lettuce by using the correct fertiliser. *Acta Horticulturae*, **29**, pp. 327–39.

Collier, G F, Huntingdon, Valerie C and Cox, E F (1979) The possible role of chlorogenic acid in calcium related disorders of vegetable crops. *Communications in Soil Science and Plant Analysis*, **10**, pp. 481–90.

Corgan, J N and Nakayama, R N (1967) Planting dates and varieties for spring lettuce. *Bull. N. Mex. Agric. Exp. Sta.* **518**, p. 10.

Corgan, J N and Cotter, D J (1971) The effects of several chemical treatments on tipburn of head lettuce. *Hort. Science*, **6**, pp. 19–20.

Crisp, P, Collier, G F and Thomas, T H (1976) The effect of boron on tipburn and auxin activity in lettuce. *Scienta Horticulturae*, **5**, pp. 215–26.

van den Ende, J (1954) Physiological diseases related to the water status in plants. *Meded. Dir. Tuinb.* (Eng Summary) **17**, pp. 615–36.

Grainger, J (1952) Damage to glasshouse crops by over-manuring. *J. Sci. Fd. Agric.* **3**, pp. 164–72.

van der Hoeven, A P, Huyskes, J A, and Rodenberg, C M (1967) Breeding for resistance to tipburn in lettuce. *Meded. Inst. Vered. Tuinbouwgew, Wageningen*, **274**, p. 8.

Jenkins, J M (1959) Brown rib of lettuce. *Proc. Amer. Soc. Hort. Sci.* **74**, pp. 587–90.

van der Kloes, L L J (1952) Tipburn in greenhouse lettuce *Meded. Dir. Tuinb.* (Eng Summary) **15**, pp. 125–39.

Kruger, N S (1966) Tipburn of lettuce in relation to calcium nutrition. *Queensland J. Agric. Anim. Sci.* **23**, pp. 379–85.

Leh, H O (1970) Investigations on tipburn of lettuce with special reference to nutrient uptake. *NachrBl. dtsch. PflSchDienst, Braunschweig*, **22**, pp. 86–9.

Olson, K C, Tibbitts, T W and Struckmeyer, B E (1967) Morphology and significance of laticifer rupture in lettuce tipburn. *Proc. Amer. Soc. Hort. Sci.* **91**, pp. 377–85.

Olson, K C (1968) Morphological studies of Packifer rupture and leaf development in lettuce. *Dis. Abst., Sect. B.*, **29**, p. 902.

Olson, K C, Tibbitts, T W and Struckmeyer, B E (1969) Leaf histogenesis in *Lactuca sativa* with emphasis upon lactifer ontogeny. *Amer. J. Bot.* **56**, pp. 1212–6.

Stewart, J K and Foster, R E (1955) Observations on rib discoloration and tipburn of lettuce. *Plant Disease Reporter*, **39**, pp. 418–20.

Thibodeau, P O and Minotti, P L (1969) The influence of calcium on the development of lettuce tipburn. *J. Amer. Soc. Hort. Sci.* **94**, pp. 372–6.

Tibbitts, T W, Struckmeyer, B S and Rao, R R (1965) Tipburn of lettuce related to release of latex. *Proc. Amer. Soc. Hort. Sci.* **86**, pp. 462–7.

Tibbitts, T W and Rao, R R (1968) Light intensity and duration in the development of lettuce tipburn. *Proc. Amer. Soc. Hort. Sci.* **93**, pp. 454–61.

Western, J H (1953) Water balance failure leads to tipburn. *Grower*, **39**, p. 983.

Lettuce-leaf deformity

Summary

Two distinct conditions are described for autumn grown lettuce. The irregular leaf development can be attributed to a range of factors including growth checks, hormone herbicide residues, molybdenum deficiency or chemical treatment.

Symptoms

A. The symptoms occur after propagation usually on the 5th true leaf upwards. Affected leaves are smaller, rough in texture with small blisters or pustules on the surface. In severe cases the leaves in-roll tightly.

B. Typical 'hormonal' symptoms occur at the 5–7 leaf stage. The leaf surface is puckered with a ragged margin and in-rolled lamina — 'tulip-leaf'. Associated with this symptom is a swollen hypocotyl with profuse aerial root growth.

Plants suffering from either symptom are usually capable of recovering and producing marketable lettuce somewhat later than unaffected plants.

Occurrence

Affected plants have been found irregularly distributed amongst good areas either singly or in small groups. Usually plants from August to December sowings are more susceptible. Symptoms show after planting and rarely in the propagation stage.

Causes

Molybdenum deficiency can induce the rolling and thin growth of lettuce leaves. Residues of hormone-type herbicides can cause distorted leaf growth, e.g. dog's tongue symptom with MCPA, 2,4–D and 2, 3, 6–TBA. The stem of the plant may elongate and adventitious roots develop at soil level. 2, 3, 6–TBA can cause the fusion of leaf lamina to form the tulip leaf type of deformity. Benomyl has been reported to have cytokinin-like properties (Skeene, 1974), which could under certain conditions perhaps upset the hormone balance of the lettuce plants. Other fungicides have similar properties.

The development of the symptoms after planting suggests that if no phytotoxic residues are present in the soil then the disorder is induced by a growth check due to either temperature, moisture availability, soil conditions or perhaps spray applications.

References

Skeene, K G M (1974) Cytokinin-like properties of the systemic fungicide Benomyl. *J. Hort. Sci.* **47**, pp. 179–82.

Martin, J A and Fletcher, J T (1972) The effects of sub-lethal doses of various herbicides on lettuce. *Weed Res.* **12**, pp. 268–71.

Tomato — blossom end rot (BER) (Plate 7.3)

Summary

The breakdown of tissue in lesions at the blossom end of tomato fruit (BER) is associated with low levels of calcium in the fruit. Restricted transpiration due either to poor root activity and consequential limited water uptake in high transpiration conditions or the maintenance of high humidity prevents adequate uptake and mobilisation of calcium to the developing fruit.

Symptoms

Blossom end rot initially appears at the end of the fruit taking on a wilted appearance. It also has been described first appearing as a grey water soaked area at the blossom end. This enlarges turns brown, and eventually develops into a dry sunken spot with a dark grey or black leathery surface. The necrotic tissue contains high amounts of phenolase. The lesions are approximately symmetrical around the blossom end and do not appear on other parts of the fruit. Lesions may appear while the fruits are very small. The distal fruit on a truss are often the worst affected.

Occurrence

BER is seldom a serious problem in commercial tomato crops in border soils. It is usually associated with crops grown in restricted volumes of rooting medium, e.g. in pots, peat modules or troughs, and where watering is inadequate. However, BER can also occur in crops grown by the Nutrient Film Technique especially early in the season when losses can be financially serious. BER is most severe at times of high water requirement during the summer.

Causes

Experiments have shown that BER is associated with low levels of calcium in the fruit. It can be induced in solution culture by omitting calcium, and a critical level of 0.08 per cent

Ca in the fruit, below which BER occurs, has been postulated by Wiersum (1966). Such concentrations indicate an important role for calcium at 'trace element' level, and contrast markedly with the far higher levels found in tomato leaves (around 2.5 per cent Ca). Movement of calcium from leaf to fruit seems to be very limited, however, and the developing fruit is primarily dependent on calcium transported directly from the roots. Cellular permeability increases with decreasing calcium content of the tissue. It is probable that the calcium deficiency causes the necrosis of the tissue in BER by causing disorganisation of membranes and organelles in the cell.

Low calcium in the fruit can be caused by one or a combination of the following factors:

Water stress

Calcium uptake and translocation is closely associated with water supply (calcium is transported in the xylem). BER is most likely to occur in periods of high transpiration, when water applications are insufficient to prevent periodic water stress. It can also occur where high humidities restrict transpiration and hence calcium transport. In such circumstances 'burning' or 'scorching' of the youngest cones may occur.

Root damage

Over-watering or high salt levels, especially if the growing medium is being kept too dry can damage roots which may prevent adequate calcium and water uptake. The incidence of BER in nutrient film culture has been associated with root death, despite adequate calcium concentrations in the solution.

Imbalance of nutrients

High K/Ca ratios and the use of ammonium–N rather than nitrate–N have both been shown to increase the occurrence of BER. Presumably potassium and ammonium ions compete with calcium ions for uptake by the plant.

Low calcium in the growing medium

This is rarely the cause of BER in crops grown in border soil. However, BER has occurred in crops grown in peat modules where it is thought the peat may have contained less than adequate levels of available calcium.

BER can often be prevented from spreading to higher trusses if, immediately an affected fruit is seen, plants are sprayed with a 0.2 per cent calcium nitrate solution at intervals of five to seven days. As a first aid measure calcium nitrate can also be included in the liquid feed but this is not possible if it is also necessary to feed phosphate due to precipitation of calcium phosphate. Therefore it is advisable to include sufficient superphosphate in the base dressing. This is also desirable because phosphate as a liquid feed is normally given as monoammonium phosphate and the ammonium nitrogen may increase the likelihood of BER occurring (see above). Careful attention must be given to watering of plants grown in peat to ensure the disorder does not occur.

References

Besford, R T (1978) Effect of potassium nutrition of three tomato varieties on incidence of blossom-end rot. *Plant and Soil*, **50**, pp. 179–91.

van der Boon, I (1973) Influence of K/Ca ratio and drought on physiological disorders of tomato. *Neth. J. agr. Sci.* **21**, pp. 56–67.

Van Goor, B J (1968) The role of Ca and cell permeability in the disease blossom-end rot of tomatoes. *Physiol. Plant.* **21**, pp. 1110–1.

Geraldson, C M (1957) Control of blossom-end rot of tomatoes. *Proc. Amer. Soc. Hort. Sci.* **69**, pp. 309–17.

Greenleaf, W H and Adams, F (1969) Genetic control of blossom-end rot tissue in tomatoes through Ca metabolism. *J. Amer. Soc. Hort. Sci.* **94**, pp. 248–50.

Hobson, G E (1967) Phenolase activity in tomato fruit in relation to growth and various ripening disorders. *J. Sci. Fd. Agric.* **18**, pp. 523–6.

Maynard, D N, Barham, W S and McCombs, C L (1957) The effect of Ca nutrition of tomatoes as related to the incidence and severity of blossom-end rot. *Proc. Amer. Soc. Hort. Sci.* **69**, pp. 618–22.

Ward, G M (1973) Causes of blossom-end rot of tomatoes based on tissue analysis. *Can. J. Plant Sci.* **53**, pp. 169–74.

Tomato — blotchy ripening (Plate 7.4)

Summary

Blotchy ripening of tomato fruit can be caused by inadequate potassium nutrition and/or environmental factors.

Symptoms

The disorder is characterised by green, yellow, or translucent areas of hard tissue on an apparently normal red fruit (Bewley and White, 1926), but the surface of the fruit is not pitted. The affected areas are, more often than not, close to the calyx (Winsor, 1960). The disorder is identical with 'cloud' in New Zealand while a more severe form is known as 'waxy patch'. On cutting open a blotchy fruit, two types of tissue are often recognisable 'brown' necrotic tissue which occurs around and between the vascular bundles and hard 'white' tissue which is often visible in the locular walls (Hobson *et al* 1977).

Causes

Blotchy ripening has been attributed to a variety of factors, both nutritional and environmental (for reviews see Cooper, 1957; Woods, 1968; Hobson and Davies, 1977). It is generally agreed, however, that the incidence of the disorder is minimised by adequate potassium fertiliser use (Winsor *et al* 1961; Winsor and Long, 1967; Ozbun *et al* 1967; van der Boon, 1973), but inadequate potassium fertiliser does not seem to account entirely for the disorder (Hobson and Davies, 1976; Hobson and Davies, 1977). A convincing explanation for the cause of blotchy ripening is still awaited. Meanwhile, high levels of soil potassium, and keeping air temperature below 29°C will help to reduce the incidence of the disorder.

References

Bewley, W F and White, H L (1926) Some nutritional disorders of the tomato. *Ann. appl. Biol.* **13**, pp. 323–38.

Cooper, A J (1957) Blotchy ripening and allied disorders of the tomato: a critical review of the literature. *Ann. Rep. Glasshouse Crops Res. Inst.* 1956, pp. 39–47.

Fletcher, J T, Harris, Patricia A and Atkinson, Kay (1976) Tomatoes diseases, pests and disorders. Yorkshire and Lancashire Region, Leeds, MAFF ADAS.

Hobson, G E and Davies, J N (1976) Protein and enzyme changes in tomato fruit in relation to blotchy ripening and potassium nutrition. *J. Sci. Fd. Agric.* **27**, pp. 15–22.

Hobson, G E and Davies, J N (1977) A review of blotchy ripening and allied disorders of the tomato, 1957–1976. *Ann. Rep. Glasshouse Crops Res. Inst. 1976*, pp. 139–47.

Hobson, G E, Davies, J N and Winsor, G W (1977) Ripening disorders of the tomato. *G.C.R.I. Growers' Bulletin*, **4**.

Kidson, E B and Stanton, D J (1953) Cloud or vascular browning in tomatoes I. Conditions affecting the incidence of cloud. *N.Z. J. Sci. Technol. Sect. A*, **34**, pp. 521–30.

Kingham, H G (1973) Other disorders. *In* 'The U.K. Tomato Manual' Ed. H G Kingham, London, Grower Books, Chapter 31.

Ozbur, J L, Boukonnet, C E, Sadik, S and Minges, P A (1967) Tomato fruit ripening. I. Effect of potassium nutrition on occurrence of white tissue. *Proc. Amer. Soc. Hort. Sci.* **91**, pp. 566–72.

van der Boon, J (1973) Influence of K/Ca ratio and drought on physiological disorders in tomatoes. *Neth. J. Agric. Sci.* **21**, pp. 56–67.

Winsor, G W (1960) A note on the grading of unevenly coloured tomato fruit from the manurial trials during 1959. *Ann. Rep. Glasshouse Crops Research Institute, 1959*, pp. 83–8.

Winsor, G W, Davies, J N and Long, M I E (1961) Liquid feeding of glasshouse tomatoes, the effects of potassium concentration on fruit quality and yield. *J. Hort. Sci.* **36**, pp. 254–67.

Winsor, G W and Long, M I E (1967) The effects of nitrogen, phosphorus, potassium, magnesium and lime on factorial combination on ripening disorders of glasshouse tomatoes. *J. Hort. Sci.* **42**, pp. 391–402.

Woods, M (1968) The effects of some nutritional and environmental factors on fruit quality in tomatoes. Proceedings 8th Congress International Institute, Brussels 1966 pp. 313–23. Berne International Potash Institute.

Tomato — bronzing and pitting (Plate 7.5)

Summary

A range of symptoms from the discoloration of the flesh of the tomato to the deformation of the skin have been attributed to infection by tomato mosaic virus (Hobson *et al* 1977).

Symptoms

A light brown coloration shows through the skin of the fruit while it is still green. The brown colour is caused by necrosis of the cells around the vascular tissue. In slight attacks, fruit show no external lesions, but on the surface of severely affected fruits irregular brownish slightly sunken lesions (pitting) are found, while the same bronzing of the tissue just below the skin can be seen on cutting the fruit. These symptoms are identical to those ascribed to the disorder 'vascular browning'.

Causes

Infection by tobacco mosaic virus is considered to be the principal cause of these fruit symptoms. The incidence, however, can be increased by unbalanced growth and poor growing conditions such as incorrect temperatures, poor light, excessive watering and nutrient levels. Losses due to virus may be avoided by the use of cultivars containing genes for tolerance to tomato mosaic virus (Darby, 1973).

References

Anon. (1972) Mosaic and streak of tomato. London, MAFF Leaflet 38.

Broadbent, L (1963) Recent work on the epidemiology of tomato mosaic virus. Proceedings 16th International Horticultural Congress. Brussels 1962. **1**, p. 104 and **2**, p. 350.

Darby, L A (1973) Tomato breeding at the Glasshouse Crops Research Institute. *Rep. Glasshouse Crops Res. Inst. 1972*, pp. 116–29.

Hobson, G E, Davies, J N and Winsor, G W (1977) Ripening disorders of tomato fruit. *GCRI Growers Bulletin*, **4**.

Jenkins, J E E, Wiggell, D and Fletcher, J T (1965) Tomato fruit bronzing. *Ann. appl. Biol.* **55**, p. 71.

Tomato — hollow (boxy) fruit (Plate 7.6)

Summary

Hollowness is associated with low light levels and a low dry matter content of fruit. It is mainly a disorder of early tomato production (Winsor, 1970).

Symptoms

Hollow fruit characteristically lack the semi-liquid tissue normally surrounding the seeds which leads to a gap between placental tissue and the outer wall of the fruit. Affected fruit are flattened and mis-shapen and are easily recognised by their ability to float in water (Winsor, 1966).

Causes

The hollow condition arises as a result of the reduced placental tissues. In severely affected fruits the outer wall of the pericarp was found to be considerably thinner than normal (Kedar and Palevitch, 1970). No single cause of hollow fruit has been established. Because the fruit on the early trusses are most frequently affected, the problem has been attributed to excessive plant vigour under poor light conditions and with insufficient potassium. The auxin-type setting hormones have been reported to increase the occurrence of mis-shapen and hollow fruits (Rilska, 1970). High levels of potassium nutrition decrease the incidence of hollowness (Hobson *et al* 1977). Fruit grown in nutrient film culture are less likely to be hollow than those grown in peat modules (Adams and Winsor, 1977).

References

Adams, P and Winsor, G W (1977) Further studies of the composition and quality of tomato fruit. *Rep. Glasshouse Crops Res. Inst. 1976*, pp. 133–8.

Hobson, G E, Davies, J N and Winsor, G W (1977) Ripening disorders of tomato fruit. Littlehampton Glasshouse Crops Research Institute. *Growers Bulletin,* **4**.

Kedar, H and Palevitch, D (1970) *Israel J. Agric. Res.* **20**, pp. 87–90.

Rilska, I (1970) *Hassadeh,* **50**(5), pp. 528–31.

Winsor, G W (1966) A note on the rapid assessment of 'boxiness' in studies of fruit quality. *Rep. Glasshouse Crops Res. Inst. 1965*, pp. 124–7.

Winsor, G W (1970) A long-term factorial study of the nutrition of glasshouse tomatoes. *In:* Fertilization of Protected Crops. Berne International Potash Institute, pp. 269–81.

Tomato — greenback (whitewall) (Plate 7.7)

Summary

Greenback is most apparent on fruit of susceptible varieties exposed to excessive sunlight. It can be avoided by growing greenback-free cultivars, using growing conditions which will avoid high air temperatures and maintaining adequate soil phosphorus and potassium levels.

Symptoms

A partial or complete ring of tissue around the calyx of the fruit which remains green (or more commonly yellow-green) while the rest of the fruit ripens normally. In modern greenback-free cultivars it is thought that white-wall tissue, an area of hard white cells along the inner parts of the locule walls, is a similar condition.

Causes

High tissue temperatures due to exposure of fruit to excessive sunlight leads to greenback in susceptible cultivars (Venter, 1970).

Low levels of phosphorus and potassium tend to aggravate the condition (Winsor, 1970) but nutritional treatments are not entirely effective (Venter, 1966).

References

Hobson, G E, Davies, J N and Winsor, G W (1977) Ripening disorders of tomato fruit. Littlehampton Glasshouse Crops Research Institute. *Growers Bulletin* 4.

Kingham, H G (1973) *In:* The U.K. Tomato Manual, Ed. H G Kingham. London, Grower Books, pp. 219–23.

Venter, F (1966) Investigation on greenback of tomatoes. *Acta Hort.* **4**, pp. 99–102.

Venter, F (1970) Beobachtungen uber du Temperatur und den Chlorophyllgehalt in Tomatenfruchten und das Aufbreten von 'Grunkragen'. *Augewandte Botanik,* **44**, pp. 263–70.

Winsor, G W (1970) A long-term factorial study of the nutrition of greenhouse tomatoes. *In:* Fertilization of Protected Crops. Berne International Potash Institute. pp. 269–81.

Tomato — flower abortion

Summary

Extremes of temperature and moisture conditions unbalanced growth and pollution of the greenhouse atmosphere can cause flower abortion.

Symptoms

Flower abortion or 'knuckling off' is preceded by yellowing of the pedicel and withering of both sepals and petals. Often large numbers of flowers drop on to the soil.

Causes

Several extreme environmental and cultural situations will increase the incidence of flower abortion; over-vegetative growth, high day temperatures, poor light and moisture stress are amongst the possible causes. Ethylene released into the atmosphere from the burning of hydrocarbon fuels will cause flower abortion. It is also suspected that the accummulation of high CO_2 levels and other pollutant gases during the night from paraffin heaters can induce this condition.

Tomato — dry set

Summary

Poor viability of pollen due to unfavourable environmental conditions prevents complete pollination, and the development of fruitlets is arrested.

Symptoms

The condition known as dry set occurs when there is no development of the fruitlet after flowering and it remains small, white and enclosed in the sepals. Fruit that partially develops to a small ripe mainly seedless fruit is known as 'chat fruit'.

Causes

Dry setting is most frequently observed on the first or second trusses of early planted crops, but can occur at other times during difficult climatic conditions. Low light conditions will reduce the viability of pollen produced, particularly by plants growing vigorously. Hot dry conditions late in the season can give rise to dry setting due to failure of the pollen to germinate on the style. Good management measures that assist pollination help towards avoiding the problem. The production of an above average number of chat fruit is also associated with unfavourable pollination conditions.

Tomato — physiological leaf roll

Summary

When there are good conditions for photosynthesis, established plants with poor setting on their lower trusses and in low night temperature situations can exhibit leaf rolling. The accummulation of starch and other carbohydrates is associated with the problem.

Symptoms

Leaves usually at the base of the plants roll inwards. The condition progresses up the plants when there are extremely unfavourable conditions.

Causes

The incidence and severity of leaf roll are reduced by shading. Removal of shoots and flower trusses causes a rapid increase in leaf roll. Copper deficiency produces leaf roll symptoms but usually there is also some veinal necrotic spotting.

High levels of sugar and starch have been found to accumulate in rolled leaves.

Flower crops

Chrysanthemum — exudation

Summary

The exudation and accumulation of salts on leaf and stem tissue of rooted chrysanthemum cuttings occur during establishment of pot plants in particular. It is considered to be a guttation phenomenon but accentuated by low humidity.

Symptoms

The exudation appears dry and powdery and usually yellowish or pale brown. The deposit is most abundant on the stem just above soil level and on the petioles and basal part of the lamina of older leaves. A deposit may be present on these areas without apparent plant damage but if the exudation extends to younger growth, severe leaf damage and death of the growing point can occur. The deposit cannot easily be brushed or washed off. Analysis has shown calcium sulphate to be the major component, with small amounts of other ions.

Occurrence

This disorder occurs in pot chrysanthemums at all times of the year, and appears in the early stages of growth after potting.

Causes

The deposition of salt is thought to be a result of guttation from the leaf, although liquid guttation droplets are seldom seen. Guttation in most crops is associated with restricted transpiration. This problem is accentuated by low humidity, and frequent overhead damping after potting is used successfully as a preventive measure. The very limited experimental work has confirmed the effect of low humidity. Deposits on the stem base are very common but appear to be of no great importance. Petiole and leaf damage is uncommon, but occasionally when it occurs can give serious losses in individual crops.

Similar symptoms have been seen in border-grown chrysanthemums, usually without serious damage, and in a number of other pot plants.

Chrysanthemum — petal necrosis (nectrolic halo)

Summary

Brown necrosis on basal petals of chrysanthemums may be related to a shortage of boron in the flower. Some fungi can cause similar symptoms.

Symptoms

Tips of basal petals show slight appearance of dry brown necrosis extending from the distal end of the petal.

Occurrence

This has been reported on Indianopolis white chrysanthemums.

Causes

Symptoms have been induced by omitting boron from plants grown in nutrient solution. When boron was removed four weeks prior to blooming the level of boron in the leaves remained adequate but the boron level in the flower was reduced. The extent of the symptoms in the flower are related to the time of removal of boron.

Somewhat similar symptoms may be caused by the petal blight fungus, *Thesonilea perplexaus* although the lesion is often elliptical and in high humidity exhibits a 'blue bloom'. Either condition can be secondarily invaded by the grey mould fungus, *Botrytis cinerea*.

Reference

Boodley, J W and Sheldrake, R (1973) Boron deficiency and petal necrosis of 'Indianapolis White' chrysanthemums. *Hort. Sci.* **8**, pp. 24–6.

Chrysanthemum — poor bud development

Summary

Low light conditions and/or copper deficiency in chrysanthemums in a peat based medium have been shown to cause poor bud development resulting in uneven flower distribution within beds or pots.

Symptoms

Flower buds fail to develop normally under environmental conditions suitable for flowering with some or all of the axillary shoots remaining vegetative, causing an uneven flower development within the bed or pot.

Occurrence

This has occurred in both pot plant and cut flower production in peat based substrates.

Causes

Exceptionally low winter light conditions and inadequate night temperatures can give rise to poor bud development. Work at Glasshouse Crops Research Institute has shown that copper nutrition can also affect bud development. Copper deficient plants show delayed flowering, loss of apical dominance and some leaf chlorosis depending on the severity of the deficiency. Of the two cultivars examined Polyanne was more sensitive than Hurricane to low copper levels. Copper is associated with the activity of several enzymes in the plant tissue including polyphenol oxidase and indole acetic acid oxidase (IAA). Flowering is probably affected by a decrease in IAA activity which in turn interferes with the regulation of the oxidation of the auxin (IAA). Exogenous application of IAA delays flowering of chrysanthemums.

Reference

Graves, C J and Sutcliffe, J F (1974) An effect of copper deficiency on the initiation and development of flower buds of chrysanthemum morifolium grown in solution culture. *Ann. of Bot.* **38**, pp. 729–38.

Chrysanthemum — hard centre (floret abortion)

Summary

The abortion of a number of disc florets in the centre of the chrysanthemum flower is thought to be caused by transient high temperatures.

Symptoms

During the development of flower buds the veil ruptures prematurely and on dissection a proportion of the disc florets are found to have aborted.

Occurrence

This disorder occurs infrequently on both pot and cut flower chrysanthemums, usually on crops timed to flower in mid to late November.

Causes

The cut flower cultivars Rivalry and the Shoesmith family have been reported as prone to this disorder. It is suggested that transient high temperatures particularly under blackout after bud initiation cause the abortion.

Chrysanthemum — excess photosynthate syndrome (EPS)

Symptoms of excess photosynthate syndrome (EPS) include bronzing, leaf roll, necrosis, thickening and yellowing of leaves.

So far there have been no advisory instances in the UK but the disorder has been reported in the American literature to affect standards mostly.

The degree of visible leaf abnormalities attributed to EPS were closely correlated with the leaf contents of glucose, sucrose, total sugars, starch and total carbohydrates. Consequently it is concluded that conditions causing high photosynthetic rates and a subsequent excessive accumulation of carbohydrates will cause EPS.

Reference

Waltz, S S and Engelhard, A W (1972) Physiological disorders of leaves of chrysanthemum cultivars relative to accumulation of excess carbohydrate. *Proc. Flor. State Hort. Soc.* **84**, pp. 370–4 Hort. Abst 1973, 6972.

Chrysanthemum — yellow strapleaf

Pale strap-like leaves developed three to four weeks after planting with retardation of new growth of axillary shoots.

The disorder is being investigated at the University of Florida where it is suspected that it is caused by an amino acid toxin produced in the soil by a micro-organism.

Reference

Anon. (1975) A new disorder in chrysanthemums. *Grower*, **83**(3), p. 125.

Carnation — calyx splitting

The calyx of the flower head ruptures causing the flower petals to droop on one side thus spoiling the quality of the flower.

The peak period for splitting appears to be March to April. Factors such as temperature fluctuations, high levels of phosphate, low nitrogen and sudden changes in climate are thought to be associated with the problem. The basic cause is a disturbance in energy balance of the flower brought about by a sudden increase in light at the time of flower opening. Splitting can be induced when the level of radiation at corolla formation is about 1.4 times the level at calyx formation. Shading minimises the risk of splitting.

Rose — proliferation after budding (Plate 8.1)

Budding takes place normally but several adventitious shoots grow out. These are apparently of bud material, but growth is very weak so the bush does not reach a saleable standard. Many cultivars can be affected, but Evelyn Fison is particularly so, and severe losses sometimes occur. Growth may be normal in the second season, but it is seldom commercially practicable to retain bushes for this period.

The condition is under investigation at GCRI where there is some suggestion that an infectious agent may be involved.

Narcissus — bull head

Very little is known about this disorder although the condition has been known for many years.

Symptoms

The disorder is found in the narcissus 'Cheerfulness' which is a double sport of the tazetta narcissus 'Elvira'. Symptoms vary according to the weather conditions. At an early stage the buds are fatter than normal and when the flowers partially open resemble drumsticks. The flower generally has a greenish coloration. In wet weather buds may rot before fully opening. Under reasonable conditions they open to a fully double flower (Cheerfulness is normally only partially double) with greenish coloration and often unbalanced shape. Under some conditions the bull-head may wither and fail to develop. When the bud is examined in detail, excessive doubling of the flower parts is found to have occurred and the normal extension of the corolla tube which occurs later in flower development (and causes the spathe to split) does not occur.

Occurrence

Fairly usual at a level of 1 to 2 per cent but stocks exist with up to 20 per cent affected. Trials at Rosewarne EHS over several years show that the condition persists in the affected plants from year to year. Hence the only control measure at present is to rogue the crop.

Causes

The cause of this disorder is unknown. It may be genetical or viral in origin. It has been suggested that drying bulbs in strong sunlight causes adherence of floral organs. Excessive hot water treatment may give similar effects.

References

Tompsett, A A (1972) Narcissus — warm storage and hot water treatment trials with the cultivar Cheerfulness. *17th Rep. Rosewarne Expt. Hort. Sta. 1971*, p. 51.

Rees, A R (1972) The growth of bulbs. London. Academic Press p. 237.

Narcissus — bud death or blindness

Symptoms

The flower fails to open, with the floral parts remaining within the spathe and becoming dry and membraneous. Similar symptoms can occur in all narcissi in the year after hot water treatment but in this case the spathe is flat and empty.

Occurrence

This condition has been known to exist for very many years in particular in the Tamar Valley of Devon/Cornwall with the *narcissus poeticus* Flore Pleno (Double White) and several other double narcissi.

The condition is only found in double flowered cultivars; the later flowering ones such as 'Double White' being particularly susceptible. Although commonly called blindness, the stems grow normally but there is a failure of the bud to develop. Incidence varies from year to year and can be as high as 50 per cent of the crop.

Causes

It is thought to be associated with checks to growth such as a change of temperature, cold nights, or hot days with dry soil conditions. Although moisture and light shade were thought to be beneficial this has not been confirmed in trials where all forms of coverage tended to increase incidence of the disorder. At the Scottish Horticultural Research Institute, planting depth did not affect the incidence.

References

Tompsett, A A (1972) *17th Rep. Rosewarne Exp. Hort. Sta. for 1971*, pp. 18–9.

McKerran, D K L and Waister, P A (1974) *Scottish Horticultural Research Institute Annual Report*, p. 27.

Rees, A R (1972) The growth of bulbs. London, Academic Press, p. 239.

Narcissus — chocolate spot

This disorder shows as chocolate brown blotches on the leaves. The blotches are 10 mm or more in diameter though elongated streaks can also occur. The symptoms appear in late spring but have little effect on growth. It is not very common with an average of one or two cases each year in the narcissus growing area of the south west, but also occurs in eastern England.

There are some cultivar differences in susceptibility. Warm wet seasons favour the disorder.

Narcissus — grassiness

Symptoms

Few flowers are produced and large numbers of leaves appear from the bulb which tends to split and produce many daughters.

Causes

Caused by a number of factors, the most likely being killing of the main shoot by bulb fly or bulb scale mite, or by too severe hot water treatment. Other possibilities are genetic instability or virus infection.

Occurrence

Fairly common in all areas but usually only a small percentage affected in any one crop.

Reference

Rees, A R (1972) The growth of bulbs. London, Academic Press, p. 236.

Narcissus — hot water treatment damage (Plate 8.2)

A range of root, leaf and flower symptoms can be caused by incorrect hot water treatment. Narcissus poeticus 'Flore Pleno' (Double White) or N. Poeticus ornatus are more susceptible than cupped varieties. If the treatment is applied too late in the development of the bulb some or all of the roots are killed causing poor growth, blotched or distorted leaves, dead buds and early senescence. Storage below 16°C before treatment can result in speckling of leaf tips, roughening and leaf distortion and flower damage ranging from blindness to reduced size split trumpets and ragged perianths.

Reference

MAF (1977) Hot water treatment of narcissus bulbs. Short term Leaflet 21.

Narcissus — bud blast

In forced crops the flower may dry up and die at a late stage, in some cases this is due to incorrect storage conditions.

Tulip — topple (Plate 8.3)

Symptoms

Occurs in forced tulips, the flower stem collapses and falls over late in flower development or even after the flowers are picked and on the way to market. A glassy water soaked area appears on the peduncle from which liquid may exude. The tissue shrinks inwards so ridges and wrinkles appear on the outer surface.

Causes

The disorder can be avoided by keeping forcing house temperatures low, but is also controlled by adding calcium to the plants' water. It has been postulated that calcium deficiency occurs locally in the rapidly elongating stem and that the nuclear and plastid membranes and plasma lemma break-down, the vacuole contents are lost and the cells die. As a result the stem loses turgidity locally and collapses.

References

Rees, A R (1972) The growth of bulbs. London, Academic Press, p. 243.
Algera, L (1968) Topple disease of tulips. *Phytopath Z.* **62**, pp. 251–61.

Tulip — bud necrosis (Plate 8.4)

Symptoms

Typically the whole stem above the top leaf including the flower becomes necrotic, and very reduced in size.

Occurrence

The disorder is most common where ventilation is poor and some bulbs are infected with *Fusarium oxysporum.*

Causes

The cause is complex, involving ethylene production and other factors. Ethylene acummulation stunts shoot growth and results in open flower development in the bulb which is easily infected with mites, fungi and bacteria producing the typical necrotic symptoms which kill the flower and shoot tip. Bulbs infected with *Fusarius oxysporum* have increased ethylene production.

References

Rees, A R (1972) The growth of bulbs. London, Academic Press, p. 225 & p. 239.

de Nunk, W J (1975) Ethylene disorders in bulbous crops during storage and glasshouse cultivation. *Acta Horticulturae*, **51**, p. 321.

Tulip — blindness (Plate 8.5)

Symptoms

In complete blindness the flower is not initiated and the bulb only contains the single leaf of a non-flowering plant. Sometimes three normal leaves are present but no flower is initiated the apex being occupied by a long narrow thong-like leaf.

More commonly flower initiation occurs normally but the flower fails to complete its development and aborts. Sometimes only part of the flower is involved, in other cases the whole shoot is killed.

Causes

The cause of failure of flower initiation is unknown but abortion is usually due to high temperatures before planting, often when the bulbs are in transit. Incorrect forcing conditions can also cause this disorder.

Reference

Rees, A R (1972) The growth of bulbs. London, Academic Press, p. 239.

Lily — bud drop

The flower buds absciss when being forced. This occurs in poor light conditions between November and February when the buds reach 1 cm diameter. It can be prevented by the use of artificial light.

Reference

Durieux, A J B (1975) *Acta Horticulturae*, **47**, pp. 237–40

Other bulb disorders

Some information on the following disorders is given in *The Growth of Bulbs,* by A R Rees (1972) London, Academic Press.

Hyacinth
  Loose bud
Iris
  Bud blast
  Bud initiation
Tulip
  Chalking
  Gumnosis
  Hard base
  Knuckling
  Watersoaking

Appendix 1   Alphabetical list of disorders

80

Appendix 2 List of colour plates

Cereals

1.1 Shrivelled grain
1.2 Seedling abnormality (propyzamide damage)
1.3 Herbicide effects
1.4 Frost damage (barley)
1.5 Frost damage (barley)
1.6 Loose ear (wheat)

Grass

2.1 Sod pulling

Root crops

Carrot

3.1 Cavity spot

Potato

3.2 Coiled sprout
3.3 Little potato
3.4 Internal rust spot
3.5 Blackening after cooking
3.6 Internal bruising
3.7 Black-heart
3.8 Second growth

Sugar beet

3.9 Strangles

Swede

3.10 Many-necking

Brassica crops

Brussels sprout

4.1 Internal browning

Cabbage

4.2 Pepper spot
4.3 Large necrotic leaf spot (Blackspot)
4.4 Vein streak
4.5 Internal tipburn

Cauliflower

4.6 Scorch
4.7 Winter stem rot (hollow stem)

Root crops and brassicas

5.1 Strangles

Fruit crops

Apple

6.1 Cox disease
6.2 Leafspot and leaf drop
6.3 Bitter pit
6.4 Cracking and russetting
6.5 Crinkle
6.6 Watercore
6.7 Carbon dioxide injury
6.8 Carbon dioxide injury (Crown heart)
6.9 Coreflush
6.10 Freezing injury (pear)
6.11 Low temperature injury (internal breakdown)

6.12 Low temperature injury (ribbon scald)
6.13 Oxygen deficiency injury
6.14 Senescent breakdown
6.15 Superficial scald

Salad crops

Lettuce

7.1 Tipburn
7.2 Tipburn (glassiness)

Tomato

7.3 Blossom end rot (BER)
7.4 Blotchy ripening
7.5 Bronzing
7.6 Boxy fruit
7.7 Greenback

Flower crops

Rose

8.1 Proliferation after budding

Narcissus

8.2 Hot water treatment damage

Tulip

8.3 Topple
8.4 Bud necrosis
8.5 Blindness

Printed in the UK for HMSO
Dd 737470 C25 1/85